PURPOSE AND DESIRE

PURPOSE

& DESIRE

WHAT MAKES SOMETHING "ALIVE" AND
WHY MODERN DARWINISM HAS FAILED TO EXPLAIN IT

J. SCOTT TURNER

HarperOne
An Imprint of HarperCollins*Publishers*

The credits on pages 319–20 constitute a continuation of this copyright page.

HarperCollins books may be purchased for educational, business, or sales
promotional use. For information, please email the Special Markets Department
at SPsales@harpercollins.com.

FIRST EDITION

Designed by Ad Librum

Library of Congress Cataloging-in-Publication Data has been applied for.

ISBN 978–0–06–265156–3

17 18 19 20 21   LSC   10 9 8 7 6 5 4 3 2 1

CONTENTS

PREFACE

Just shoot me now . . .

On one side is the ebullient and engaging Will Provine[*] telling me:

> There are no gods, no purpose, no goal-directed forces of any kind. . . . There is no ultimate foundation for ethics, no ultimate meaning to life, and no free will for humans, either.[*]

On the other side, the astonishing Francis Crick says to me that

> [science has shown you that] your joys and your sorrows, your memories and your ambitions, your sense of identity and free will, are in fact no more than the behaviour of a vast assembly of nerve cells and their associated molecules. As Lewis Carroll's Alice might have phrased it: "You're nothing but a pack of neurons."[1]

---

[*] From Provine's interview in the 2008 documentary on intelligent design theory *Expelled: No Intelligence Allowed*, starring Ben Stein. Sadly, Provine died in 2015 after a long struggle with cancer.

Standing in front of me, the relentless Richard Dawkins informs me,

> The universe we observe has precisely the properties we should expect if there is, at bottom, no design, no purpose, no evil, no good, nothing but blind, pitiless indifference.[2]

And, standing behind me, the dangerous Daniel Dennett hectors me through his magnificent beard that it's all a mindless algorithm:

> No matter how impressive the products of an algorithm [i.e., natural selection], the underlying process always consists of nothing but a series of individually mindless steps succeeding each other without the help of any intelligent supervision.[3]

Beset from all sides, the message could not be more clear. It's all endless, mindless, purposeless, relentless, no point, no purpose, no intent, no goal, no . . . nothing.

Oh, dear!

I'm indulging in a little bit of caricature here, of course. Provine, Crick, Dawkins, and Dennett often show up as exemplary bêtes noires on creationist, intelligent design, and other generally anti-Darwinian websites showing us what the dangerous Darwinian world has in store for us, as heralded by these supposed Four Horsemen of the Evocalypse. It's an unfair accusation; all these authors have written extensively, deeply, and subtly on the central paradox of Darwinism: the problem of what seems like design, intentionality, intelligence, and purposefulness in the living world. One cannot read their works (or meet them in person) without sensing in them the joy that comes from being able to frolic and wriggle through the tangled banks of the Darwinian idea.

Nevertheless, one also gets the impression of a different form of glee—an almost Calvinist glee—at the prospect of using Darwinism as a cudgel to dethrone what Carl Sagan once described as the demon-haunted world[4]—a world that is permeated with spirits and gods that shape and control it, intelligently, purposefully, and intentionally. Spending time wandering through the Darwinian thickets with our four horsemen and their acolytes, one is left wondering whether their philosophical joie de vivre has not lapsed into inebriation and intoxication—and into intoxication's usual end point: grumbling nihilism and meaninglessness.

One could dismiss this type of talk as the gloom-mongering old men like me are prone to, but I don't think that's all it is. An anecdote: a few years ago, I was sitting at lunch at my university with a friend and colleague along with a small gaggle of students. As we often did on these occasions (as we are both professors and therefore pedantic), we were putting on a little bit of a performance for our lunch companions, our discussion turning on the very problem of intentionality and purposefulness that the four horsemen had denounced most forcefully. Our little debate centered on whether intentionality is something "real," that is, a genuine phenomenon that exists in and of itself, and is therefore susceptible to scientific inquiry; or whether intentionality is only "apparent," that is, mindless molecular churning that creates only an *appearance* of intentionality in the credulous human mind. At one point, I asked my friend the question: What if intentionality *is* real? Furthermore, what if intentionality is not only real, but is actually the most important attribute of life? Could we then be scientists true to our calling if we ignored it? His answer was no, we could not then be scientists; we would have to be something else—philosophers, perhaps, or theologians. That unsatisfying end was where we left it that day, with my questions unanswered. So I am left with them: What if the distinctive attributes of life truly *are* intentionality, purposefulness, and the wants and desires, however inchoate, of living things? How could we possibly understand life if we deliberately ignored those attributes?

And another anecdote: this book started out as a sequel to my book *The Tinkerer's Accomplice*, which explored what I thought then to be the unsolved and very real problem of biological design. I had included in that book a short chapter on intentionality, arguing that intentionality was not some magical thing; it was in fact the obverse of cognition, which itself found its origins in the homeostasis of the brain "ecosystem." I thought it was a pretty little idea, not into the weeds at all, but my chapter on intentionality provoked emphatic displeasure from a reviewer writing for the books section of a prominent newspaper. This reviewer was a fellow scientist of some renown, and his displeasure was provoked not so much by what I had to say about intentionality, but by the fact that I had had the temerity to even bring up the subject, which he felt had no place in a book that presumed to be scientific. Fair enough—reasonable people can differ on this. But he had gone on to demand that I should have made known my religious beliefs (I'm a Christian, if that matters, albeit not a very good one). This struck me as strange, even a tad illiberal. Would he have demanded a religious confession from an author whose views he had found more congenial? And what would he have done armed with my confession in hand? You see what happened here: I had raised issues as a scientist trying to explain phenomena in my field of study, but my reflections were seen as being so out of bounds (and threatening, presumably) that they could only be explained by religious motivations that I was trying to sneak into my work. I had clearly struck a nerve.

The conclusion I draw from these two anecdotes is that we scientists have a problem with the very idea of purposeful life, and the problem is widespread. To be scientists, it seems that the first thing we must do is put on blinders to squishy ideas like purposefulness, intentionality, and a guiding intelligence. But what if these ideas are central to the phenomenon we are studying—life itself? Can questions of purpose and desire be imposed only from outside the work of science? Isn't that precisely the opposite of what we, as scientists, should do? Yet, we force ourselves into a Hobson's choice on the matter: accept intentionality and purposefulness as real attributes of life, which disqualifies you as a scientist, or become a scientist and dismiss life's distinctive quality from your thinking.

———•———

Which brings me to the book you hold in your hand. I have come to believe that there is something presently wrong with how we scientists think about life, its existence, its origins, and its evolution. It's bad enough that we are somehow forced into making the Hobson's choice described above. What's worse is that being forced to make the choice actually stands in the way of our having a fully coherent theory of life, in all its aspects, most notably its evolution. In other words, this bias is now hindering scientific progress.

This is tragic, because there is nothing in the nature of the science of life per se that forces the choice to be made. The problem traces its beginning to the early twentieth century, when biology sold its soul, so to speak, committing the practitioners of the science of life wholesale to the essentially philosophical premises of mechanism and materialism. At the time, this probably seemed a sensible thing to do, in no small part because biology (and evolutionism) had wandered off into a muddle of obscurantism. Turning away from that older tradition to take a new cold hard look at the material and mechanistic underpinnings of life brought much-needed clarity to our understanding of the phenomenon of life. That is why much of the twentieth century is rightly considered a golden age for the science of life.

But biology paid a price for that clarity, because along the way, it lost its hold on any claim that it was a *distinct* science of life. Biology could be a distinct science, after all, if and only if life itself were a distinctive phenomenon. Yet lurking in the fine print of that Faustian bargain struck then was the premise that life is not distinguishable in any ultimate way from the rest of the cold and impersonal universe we inhabit. Glossing over this clause (or, it must be said, the enthusiastic embracing of it) has drawn modern biology into a bit of a philosophical pickle, to wit: if biology claims to be a distinct science, on what grounds is the distinction built? For much of the twentieth century, we could afford to ignore this question, or at least not to engage it too critically, as we gathered up the pretty scientific baubles that lay newly strewn about us. We can ignore it no more, because leaving the question unresolved has brought biology to the brink of a philosophical and scientific crisis.

Let me frame the crisis for you this way. Imagine a cumulus cloud and a cauliflower, and answer the question, Which one is alive? I expect no one will have any trouble identifying the cauliflower as alive and the cumulus cloud as not, but how do we know? There are many similarities between them, after all. When I am in a flippant mood, I point to the fact that both are white, and both are puffy, so what's the difference? Putting that flippancy aside, the similarities between cauliflower and cumulus cloud actually go quite deep.

For example, both are what we call *open thermodynamic systems,* that is, organized streams of matter and energy that, through what has come to be called the Fourth Law of Thermodynamics, generate a peculiar and specified orderliness. The cumulus (pillow) cloud can be named as such largely because this spontaneous orderliness is manifest as puffy (turbulent, to be precise) clouds of condensing water droplets. It is an organized stream of matter and energy that has qualities distinctive enough to warrant a name. The cauliflower does this too—there is really no such *thing* as a cauliflower; it is as transient in its own way as a cloud on a summer's day. So there is a deep thermodynamic similarity between the two.

We could try to draw the distinction based upon the sources of energy driving each system. The cumulus cloud relies on a continuous input of energy from the sun, and when the sun departs from the sky, so too does the cumulus cloud. The cauliflower does this too, of course, but there's a complicated suite of biochemical transactions interposed between sunbeam and cauliflower that does not exist in the cumulus cloud. The cauliflower captures light energy in glucose, whereas the cumulus cloud captures light energy in heated air and water vapor. These are really distinctions only in order and not in kind, however, and it is distinctions of kind that we seek. Deep down, the cauliflower is still an open thermodynamic system that does not stand out in any fundamental way from the open thermodynamic system of the cumulus cloud. The differing complicated details do not, in the end, allow a fundamental distinction to be drawn.

One could argue that the distinction lies in the selection process (natural or artificial) that has produced the unique collection of genes that specify the cauliflower. It's an interesting point. *Brassica oleracea,* the formal species name for what we call cauliflower, exists today because it is the descendant of a long legacy of sorting of hereditary memories that "worked"—that is, hereditary memories that enabled the plants housing them to reproduce. Those memories are encoded in the long strings of nucleic acids that are carried in the precious purse of the nucleus. In short, cauliflowers have genes, whereas cumulus clouds obviously do not: distinction drawn and done. But let's not be so fast, because there's a deeper question that first has to be addressed before we can declare the matter done, and that is, What exactly do we mean by "hereditary memory"?

If we dissociate memory itself from the mechanisms and materials that encode it, like DNA, we are led to a fundamental definition of memory as some means of shaping the future as a reflection of the past. When we do so, we arrive at the startling conclusion that cumulus clouds also have a sort of hereditary memory. Cumulus clouds don't really exist in and of themselves; they are the visible manifestation of a locally constrained water cycle. They need both heat and humidity to drive them, and so they tend to form over places where there is a ready source of water for evaporation. Often that water has come as rainfalls from fully mature cumulus clouds. That this could be a kind of hereditary memory may not be immediately obvious to those of us who live in moist environments, like the northeastern United States, but it is brought starkly into relief in dry habitats, like the semitropical arid savannas and deserts of Namibia, where I have spent a lot of time watching clouds.

In that country, there are no permanent rivers. Rather, the land is laddered with "linear oases," ephemeral rivers that drain west from the high plains through the Great Escarpment and coastal plains to the Atlantic Ocean (although it must be said that the water in them almost never actually gets to the ocean). During the humid summers, lines of cumulus clouds frequently form over these linear oases, eventually dropping water right back into the comparatively moist soils whence it came. Thus the action of yesterday's locally constrained water cycle sets up the conditions to power tomorrow's turn of the water cycle crank. We have therefore a kind of hereditary memory at work. So, if it's to memory we're looking to draw the distinction between cumulus cloud and cauliflower, the ground turns out to be a little shaky there, too, provided we are willing to dig a little. Again, we fail to draw a distinction in kind.

If not thermodynamics, energetics, or heredity, where then should life's distinctive attribute be sought? I argue in this book that *agency* is where the distinction can be reliably drawn. The cauliflower can be construed as a collection of intention-driven *agents* that go about purposefully building a cauliflower. Pay close attention to the wording I used. It's difficult to imagine such agency in a cumulus cloud. We can make complete sense of the cumulus cloud as a physical interplay of heat, moisture, and gravity, but it would strain credulity to go on from there to say that the cumulus cloud exists because it *wants* to exist. To say that a cauliflower exists because, in some deep sense, it *wants* to exist might strain credulity as well, but with a squint and a hope, we can almost see how it could be true.

In a nutshell, this is where the crisis of biology looms, because our prevailing modes of thinking about life—the triumphant confluence of mechanism, materialism, and atomism that has made the twentieth century a golden age for biology—do not deal well with the concept of agency: that ineffable striving of living things to *become* something. This was the real source of annoyance for that reviewer of *The Tinkerer's Accomplice*, because intentionality is, among other things, the manifestation of living agency, as are purposefulness, desire, and striving. Philosophically fencing those things off, as much of the modern science of biology has done, has brought the science of life to the brink of its crisis.

I use the word "crisis" deliberately, because there is more at stake than quibbles and mind play over clouds and cauliflowers. What is at stake is whether there will be a coherent theory of life. Without a coherent theory of life, whatever we think about life doesn't hold water. This applies to the major contribution we claim that the modern science of life offers to the popular culture: Darwinism.

It is from the fount of Darwinism that the Four Horsemen of the Evocalypse draw sustenance, and it is through an appeal to the scientific validity of Darwinism that their acolytes make their bid to set the terms of our modern culture. Yet there sits at the heart of modern Darwinism an unresolved tautology that undermines its validity. We scientists might not be troubled by this, but we should be, not least because the failure to recognize it closes off modern evolutionism from many big problems it should be capable of answering: the origin of life, the origin of the gene, biological design, and the origins of cognition and consciousness, to name a few. Intentionality and purposefulness are important to all these unresolved big questions, and yet we are very quick to fence these off behind a wall of denial. Instead of a frank acknowledgment of purposefulness, intentionality, intelligence, and design, we refer to "apparent" design, "apparent" intentionality, "apparent" intelligence.

I deal more fully with all of these things in the chapters that follow, but I want to emphasize the central question that motivates this book: can we construct a credible and coherent theory of life that would obviate the Hobson's choice that modern biology has imposed upon itself? And if so, what would such a theory look like? I claim that there is such a theory and that the path that will lead us out of the Hobson's choice is a proper understanding of the phenomenon of *homeostasis*. Keep that term in mind, because it is a profound concept to which we shall return again and again throughout this book. It is also a dangerous concept because following its logic invites us to look anew at ideas and philosophical positions that have long been regarded as scientific heresy. Just warning you. It is my hope, though, that this little venture into heterodoxy will be worth it, and that you, gentle reader, will find it worthwhile to come along.

1

THE PONY UNDER THE TREE

I've never been able to stand the sight of blood. Even the thought of it sends a cold shiver up my spine.

This is an unfortunate handicap for a physiologist like me. As it is mostly practiced today, physiology is a medical discipline. Being a research physiologist therefore usually means studying people, or animals that are similar enough to people so that what we learn from them might be useful for people. Studying people is obviously out for me: the involuntary look of horror that sweeps over my face at hearing of someone's injuries, even fake injuries depicted in a television show, would alarm my patients and immediately disqualify me. Nor am I any better when it comes to the animal proxies physiologists commonly use: dogs are some of my favorite people; cats intrigue me with their airs of exotic disdain; rats and mice reveal wonderful hidden personalities once they decide you are not trying to eat them. Don't even mention chimpanzees.

Fortunately, I've always found some way to work around my squeamishness. I studied alligators for a time. I rationalized this by an appeal to fairness: they could turn the tables and eat *me*, after all, and without a shred of remorse. Birds' eggs (or more precisely, birds' embryos living in eggs) sufficed for a while: I suppose it was easier to work on something that couldn't look me in the eye, at least as long as I didn't allow it to hatch. Finally, there were termites, whose built structures fascinated me and which live in societies where life is cheap and the massacre of multitudes is routine. What would it matter if a few thousand more were sacrificed on the altar of science?

And that is the path that led me one day to a grassy field in northern Namibia, about to do diesel-powered surgery on two termite mounds.

It started off as a whim. The previous evening, my friend Eugene Marais and I were sitting outside at the research farm where I do my work, enjoying a beer—well, maybe more than one. We had been observing the termites' ability to repair small wounds inflicted on their mound—a hole drilled in the side, that sort of thing. Somehow these creatures—stupid, blind, and clumsy as individuals—knew en masse precisely where damage had been done and could mobilize a horde of nest mates to repair the damage, quickly and efficiently. We wanted to know how they pulled it off.

I don't remember who had the idea first, but the gist of our brainstorm was to quit fooling around with little hurts to the mound. Rather, we wondered what would happen if we just lopped the whole thing off? "Complete moundectomy" was our flippant term for the procedure we were contemplating. And because these were termites, we could act on our whim—glorious freedom!—safe from the bureaucratic killjoys who would have to be given a say if we were studying a creature that had a backbone.

So we got busy the next morning selecting and preparing a couple of mounds for our frivolous experiment. The day after that, we brought in a front-end loader and scraped both mounds right off at the ground. Because these termites' colonies are located underground, deep below the level where the blade of our machine would cut, the "patients" survived the procedure just fine and immediately began to rebuild their mounds. For the next ninety days or so, we visited our subjects daily and photographed their progress, always from the same vantage point. When our time in the field was at an end, we lined all the photographs up in my computer and made a time-lapse movie with them.

We had a rough idea what our movie would show. The mounds had regrown to about their original size and shape, even reproducing the mounds' prominent northward tilt. That was not a surprise, although how quickly the termites did it was—about three months for a construction project that involved gathering and moving about three tons of soil up into the growing mound. But the time-lapse video revealed a dimension to the reconstruction that could not quite be captured by beginning-and-end comparisons. To me, these mounds were *alive:* pulsing, heaving themselves up from the mud like hideous beasts; sometimes like cockroaches, shying away from the intense sun; sometimes aiming defiantly right at the sun, as a moth flies to a flame. These were no mere piled-up heaps of dirt; they were *beings* that knew what they were supposed to be and were intent on becoming it.

But attributing sentience or intentionality to such a thing is an impure thought in modern biology. Intentionality is subjective, vague, and unquantifiable, making it dangerously easy to project your own persona onto what you are observing. This is why modern scientists are trained from conception to view the world from a stance as far removed from sentience as possible, to explain what they see as the product of the relentless flight of multitudes of atoms (or termites) hurtling willy-nilly. If you have to fall back on fuzzy ideas like intentionality, that's prima facie evidence that you're not thinking about it hard enough: real scientists do it reductively.

There's virtue in this, of course. Scientists are as prone to flights of fancy as anyone, and this cold-eyed reductionism usually serves us well as a check on our natural exuberance. Virtue can be *too* scrupulously applied, however, and when it is, well . . . to paraphrase Mr. Bumble, science, like the law, can be idiotic.*

And this is the dilemma those time-lapse videos presented: the mounds certainly *looked* alive, but *were* they alive? The reductive reflex is to say, "Of course the mounds weren't alive, how could they be?" For one thing, none of the criteria we usually apply to distinguish the quick from the dead seem to fit. The mound has no genes, it doesn't reproduce, and it doesn't breathe. No matter how complex its structure and hypnotic its form, it's still just a pile of dirt, no more alive than a cloud in the sky.

But the impression from those time-lapse videos *was* compelling, and there comes a point for all scientists when they have to decide whether to believe the plain evidence of the senses, even as every lesson of their scientific training admonishes them not to. To properly judge the matter, we must therefore look critically at our criteria for deciding whether such a thing as a termite mound is alive or not.

Certainly the termites that build the mound qualify as alive, but what precisely qualifies them so? The usual criteria we biologists invoke are genes, reproduction, and metabolism. So, which apply to termites? Genes and metabolism the termites certainly have, but the workers that build the mound do not reproduce, leaving that chore to their fecund mother, the queen. So, if the workers score two out of three, are the workers then two-thirds alive, leaving only the queen to taste life's full measure?

Blithe sophistry, you say? Well, how about just metabolism, then? That certainly seems to be a distinctive trait of life, which even the worker termites have in full: they breathe, they eat, they keep their bodies in good trim, they avoid inevitable death as best they can. Surely the mound cannot lay claim to *that*?

Again, let's not be so quick to leap to what might seem the obvious

---

* From Chapter 61 in Charles Dickens's 1838 novel *Oliver Twist*.

conclusion, because we first need to ask, What precisely do we mean by metabolism? Here, we run into a little problem of language: a termite is not an *is* as much as it is a *does*. It is a highly ordered stream of matter and energy, which takes the ephemeral form of a termite, that is continually built up as fast as it breaks down.

What about the mound? Well, that too is more of a verb than a noun. The mound is highly ordered and dynamic—rain and wind strips it of nearly a quarter ton of soil each year, which the termites must assiduously replace, grain by grain, just as the myriad cells in a termite's body must do with every atom that flows through them. And the soil grains are replaced in a highly specified way that sustains the mound's ephemeral form, if not its actual substance. What, then, is the difference between termite and mound? At this fundamental level, it's hard to see.

If we can't rule metabolism out of bounds for the mound, that leaves genes as the essential disqualifier of the mound as alive. I'll have much more to say about genes later, but for now I'll just assert that the matter, like metabolism, is actually not so clear-cut. All living things, as far as we know, possess some form of hereditary memory, that is, some way of replicating in the future what was successful in the past. Genes, by virtue of their ubiquity, are marked as a very (and that is understating the case) successful form of hereditary memory. So successful is this form of hereditary memory that every living thing on Earth uses it. This does not preclude the existence of other forms of hereditary memory, however.

Let me illustrate using the supposedly non-living termite mound. The mound far outlasts the lifetime of any worker that builds it: a mound persists for the lifetime of the queen, more than a decade, while each worker has a life expectancy of a couple of months at best. If a colony dies and a new colony takes up residence in the vacated mound, the mound can even, at odds of about one in four, persist many years more. In that sense, a mound is the bequest of one generation of workers (or one colony, no longer living) to a subsequent generation. The mound, therefore, is as much a hereditary legacy to the termites occupying it as the strands of DNA inherited from their parents. The context, form, and longevity of the memories clearly differ, but the mound embodies hereditary memory all the same.

So, what is the best guide for answering what should be a simple question: is this animated heap of soil alive or isn't it? Do we trust our cool reason, which says that it is not, or our emphatic intuition, which says that it is? As we have just seen, cool reason seems only to lead us into a muddle, leaving the seemingly rational view—that the mound is clearly not alive—exposed as more prejudice than truth. To stick with the rational view is in fact to fall back on an article of faith—that even inconclusive reason is better than the emphatic intuition that the mound is indeed alive.

———•———

This uncomfortable spot is the starting point for the broad question that will be my theme throughout this book: do we have a coherent theory of evolution? The firmly settled answer to this question is supposed to be "yes," because Charles Darwin gave us a way to apply cool reason to answering life's mystery of mysteries—the evolution of life in its "endless forms most beautiful," even unto beings that can contemplate their own origins. Now I want to be very careful how I say what I next wish to say, because it could easily be misconstrued. I intend to argue in this book that the answer to my question might actually be "no."

That qualified "no" may lump me together, in the eyes of some, with others identified as anti-Darwinist. I can only plead that the accusation is unfounded. Darwinism is an idea of intoxicating beauty, and yet there has been for many years a muddle at the heart of it, at least in its modern form. For the most part, we've safely been able to ignore that muddle because Darwin's idea explained so much that it has kept us happily occupied at figuring out what it *did* explain. We're now coming to the point, though, where what it *cannot* explain is coming into stark relief, making it impossible any longer to ignore the muddle.

For example, we don't have a good Darwinian explanation for the origin of life. Part of the reason for this is that we don't have a good Darwinian explanation for what life is in the first place. Nor do we have a good explanation for the origin and evolution of the cornerstone of the edifice of modern Darwinism, the gene. If that weren't enough, Darwinism is also having a rather hard time explaining what an organism *is,* or why life seems to tend toward "organism-like" assemblages, or as I have argued in another book, why living things are actually (not apparently) well-designed. For the longest time, we've been able to fudge these problems, carried along on the faith that, to paraphrase the punch line of an old joke, there had to be a pony in there somewhere.* But the dread possibility is beginning to rear its head: what if the pony isn't there?

The problem for modern Darwinism is, I argue, that we lack a coherent theory of the core Darwinian concept of *adaptation.* As the conventional story goes, adaptation is the "good fit" between organism and environment, that suite of behaviors, attributes, phenotypes, whatever we wish to call them, that enable "fit" organisms to be more fecund than organisms that are not so "fit." This idea, so brilliantly simple that Thomas Huxley rebuked himself for his own stupidity at not seeing it before his friend Darwin pointed it

---

* There once was a family so poor that one Christmas they could afford to put only horse manure under their Christmas tree. The next morning, their little daughter saw the pile of manure and squealed with delight. When her puzzled parents asked why she was so happy to see a pile of horse manure under the tree, she clapped her hands together and chirped, "Because I know there's a pony in there somewhere!"

out,* is *the* core operating principle, pure and simple, of the theory of evolution by natural selection. If adaptation does not work, natural selection does not work, period.

It follows that we should therefore have a pretty good idea, commensurate with our confidence, of what adaptation is. In reality, our conception of adaptation rests on a very shaky foundation. To illustrate, consider how a recent (and admirable) textbook of evolution put it: "Adaptations are the products of natural selection, while adaptation is the response to natural selection."[1] This demonstrates, in one short and elegantly crafted sentence, The Problem: our current conception of this core evolutionary idea is essentially meaningless. What is adaptation? The product of natural selection! What is natural selection? The outcome of adaptation!

This type of reasoning is known formally as a *tautology*, which ordinarily is ranked as one of the elementary logical fallacies, an argument wherein the conclusion is a restatement of the premise, for example, "it is what it is." Yet there it is, resplendent on the page, as it has been on perhaps hundreds of other pages over the past century and a half. For Darwinism to make sense (and I want deeply for it to make sense), the tautology somehow needs to be resolved.

But how? One is tempted to say that more research will clear it up eventually, given enough time, brains, and money to fund the quest—in short, to keep looking for that pony under the tree. I'm doubtful that will work, however, for the obstacle to resolving the tautology is not that we know too little—far from it—but that we aren't thinking properly about what we do know. In short, the obstacle is largely philosophical, and the stumbling block is the frank purposefulness that is inherent in the phenomenon of adaptation.

To see adaptation at work, which is something we physiologists

---

* The self-rebuke, uttered when Huxley first read Darwin's manuscript of *On the Origin of Species*, was, "How extremely stupid of me not to have thought of that!"

spend a lot of our time doing, is to witness a phenomenon rife with purpose, intentionality, and striving. Such things do not sit well in the modern metaphysics of biology, which regards squishy things like purposefulness, intentionality, and consciousness as somehow illusory. For a biologist to treat them as something real is therefore a modern heresy, and those who advocate it suffer the fate that heretics often do: they are cast mercilessly from the altar.*

Yet disposing of a heretic does not make the uncomfortable question go away. And the uncomfortable question is this: what if phenomena like intentionality, purpose, and design are not illusions, but are quite real—are in fact the central attributes of life? How can we have a coherent theory of life that tries to shunt these phenomena to the side? And if we don't have a coherent theory of life, how can we have a coherent theory of evolution? This is the hard nut that has to be cracked, and this leads me to the other theme I develop throughout this book: the hammer that will crack the nut of a coherent theory of life will be the largely misunderstood, widely trivialized, but profound concept of *homeostasis:* what I call Biology's Second Law.

---

* Consider the harsh reception of Thomas Nagel and his 2012 book criticizing the Neo-Darwinist "consensus," *Mind and Cosmos: Why the Materialist Neo-Darwinian Conception of Nature Is Almost Certainly False.*

2

Biology's Second Law

Are there scientific laws in biology, something akin to the First Law of Thermodynamics (conservation of energy) or to Newton's First Law of Motion (conservation of momentum)? If there is any claim to Biology's *First* Law, the Darwinian idea—evolution by natural selection—has a very strong one. As far as most biologists are concerned, that's as far as it needs to go. In *Darwin's Dangerous Idea* Daniel Dennett expressed this in a penetrating phrase: all problems concerning life, be they scientific, philosophical, or historical, should be etched into clear relief under Darwinism's "universal acid."* If this is true, then biology simply has no need for a *Second* Law. Of course, not *all* problems seem to have yielded to Darwinism's universal acid—I mentioned some recalcitrant ones in Chapter 1 and will return to

---

* "Universal acid" is described by Dennett as a boyhood fantasy, an acid that is so corrosive that nothing can contain it. As Dennett describes it, "[universal acid] dissolves glass bottles and stainless-steel canisters as readily as paper bags. What would happen if you somehow came upon or created a dollop of universal acid? Would the whole planet eventually be destroyed? What would it leave in its wake? After everything had been transformed by its encounter with universal acid, what would the world look like?" (*Darwin's Dangerous Idea*, p. 63).

them in subsequent chapters—but let's leave those for now and try to flesh out what I claim Biology's Second Law should be.

It has a handy and well-known name—homeostasis*—which is usually dryly defined as "a state of internal constancy that is maintained as a result of active regulatory processes."[1] The concept is a central one for the science of physiology, of how living things *work*, to borrow from one author's pithy description.[2] Our own body temperatures are tightly regulated, for example, and so we speak of body temperature homeostasis, literally the steady (*homeo-*) state (*-stasis*) of our body's temperature, and the mechanisms that underlie the body's "thermostat." The same can be said for a myriad of other functions, such as regulation of water in the body (water homeostasis), of salts (salt homeostasis), and of the blood's acidity (blood acid-base homeostasis).

This rather anodyne description of homeostasis doesn't really do the concept justice, though, because homeostasis stands today as one of the least understood and most widely trivialized concepts in modern biology. This seems at first to be a startling thing to say. If homeostasis is the foundational principle of physiology, we should have a pretty good idea of what it means. But we don't, and the obstacle is clear: it's *mechanism*. When physiologists speak of homeostasis, their reflex is to delve deeply into the delicious details of mechanism—of, say, the body's "thermostat," as if the body were a house controlled by a machine. In other words, to study homeostasis is to dig into the realm of the mechanic, which, by definition, is concerned primarily with *machines*.

This invites the question: is life in fact a machine? Spend any time with something living, and you will see instantly that the answer is both yes and no. Mechanism is at work, to be sure, fascinating mechanism in fact, and it is essential on many levels—for the sake of

---

* The word "homeostasis" was coined by the American physiologist Walter B. Cannon in the 1920s.

innate curiosity, to develop better medical therapies, to name two—
that we understand life on those terms. Yet striving and desiring
seem to occupy a lot of what life does. My dog strives to tear his
chew toys apart into little pieces, seems to relish it, and devotes a
great deal of his day to the task; how does that *work*? A fish strives to
find shelter or lunch or love; how does *that* work? Even single-celled
creatures will seek light or shun it. What meaning should we ascribe
to these *desires,* to use a loaded word?

It is one thing to say that there is an intelligible sequence of connected events that leads living things to *appear* to strive, just as
in a machine, cogs turn axles that turn other cogs, ad infinitum.
But that appearance of striving in no way negates the quality—the
reality—of the striving of something that is alive. Machines might
appear to strive, they may come unnervingly close to actually striving, but it strains credulity to assert that a machine can ever *want*
anything, to have desire. Indeed, it actually seems more straightforward simply to acknowledge life's *actual* striving and desire,
compared to some of the tortuous reasoning that today goes into
denying it.

The stumbling block, of course, is that it seems hard to attribute
striving and desire to anything living without getting into un*scientific* ways of thinking. How do you do an experiment with desire?
To be sure, you can tweak chemicals in the brain and body or tickle
a nerve cell here and there and see what happens. But, in doing so,
have you really gotten to the nub of *what* striving and desire *are*?
In short, we find ourselves at the Hobson's choice I introduced in
the Preface: either acknowledge striving and desire as real but sacrifice being a scientist, or become a scientist and deny the reality,
independent of mechanism, of striving and desire. That's a devil's
bargain we don't have to strike, though, because there actually is a
rational explanation for striving and desire, and the font of that living desire is this strange and misunderstood concept of homeostasis.

This makes homeostasis a very subversive idea, quite unlike today's tamed and neutered version of it.

———•———

The credit for the concept of homeostasis goes to Charles Darwin's French contemporary, the great physiologist Claude Bernard (1813–1878). I will have more to say about Bernard momentarily, but for now let us simply state the matter as he did, in a famous aphorism from his 1865 master work *Introduction à l'étude de la médecine expérimentale (Introduction to the Study of Experimental Medicine)*: "La fixité du milieu intérieur est la condition d'une vie libre et indépendante." Or, in English, "The constancy of the internal environment is the condition for a free and independent life."

To approach homeostasis as a problem of mechanism, as we do today, you necessarily see it as an outcome of a type of causation: how event A (say, reduced body temperature) leads to event B (sensing the change of body temperature) and hence to event C (shivering to heat the body) and ... back to A (restored body temperature). If you parse Bernard's aphorism carefully, though, you can see that was not really the meaning Bernard intended to convey.

Bernard did not phrase his aphorism as an invitation to discover a *mechanism* of life. Rather, his aphorism is a statement of the *nature* of life. Specifically, homeostasis is not the *outcome* of life; it is life's *antecedent,* and the question that Bernard is really inviting us to explore is more the "why" rather than the "how": *Why* should body temperature be regulated? Not *how* is it regulated? Bernard's focus on the "why" is what makes homeostasis as profound an idea as Darwin's, which was also, at root, a theory of "why."

"Why" questions can be deeply unsettling, for within them lurks subtlety and beauty, and perilous truth. This is why Daniel Dennett famously characterized natural selection as a "dangerous idea."[3] Its "how" dimension is simple almost to the point of triviality, but its "why" dimension undermined most of the comfortable assumptions that prevailed in Darwin's day about life, its purpose, and its past. In fact, it undermined them so completely that we now inhabit an entirely new intellectual landscape, one that is so comfortable to us today that we scarcely notice anything strange about it. But it is a landscape that would be quite alien to most of Darwin's contemporaries.

So it is with homeostasis. Its operation—its "how" dimension—is mere mechanism, but its implications for the "why" of life—Bernard's main point, after all—is deeply subversive of the comfortable assumptions about mechanistic life that presently reign. It is, arguably, "Bernard's dangerous idea," the tonic that quenches Darwinism's universal acid before it cuts too deeply. To fully understand life in all its manifestations—its function, its evolution, its origin—we need to unpack the wonderful subtlety of homeostasis, as Bernard himself intended.

———◆———

We can illuminate the strangeness of homeostasis with a little thought experiment. In southern Namibia, near the tiny town of Asab, there stood for many millions of years a famous rock formation, commonly known as the "Finger of God,"* although it was officially known by its more descriptive Nama† name, the Mukurob,

---

* [−25.497600o, 18.171285o], in case you want to look it up.

† Nama is one of a large family of so-called click languages spoken by the many Khoekhoe (pronounced "kway-kway") tribes of southern Africa. These include the Nama, Damara, and Hai‖om (or Bushmen, to use the more familiar, if now disfavored, designation). "Mukurob" is a Europeanized rendering of the actual pronunciation, which in standard orthography is rendered "Mû!kharub," with the circumflex (^) indicating a nasalization of the "u" and the exclamation point representing the so-called alveolar click, where the tongue is pulled down from the roof of the mouth to make a hollow "pop" sound. The diversity of clicks in the Khoekhoe language family (more formally known as Khoekhoegowab) is immense.

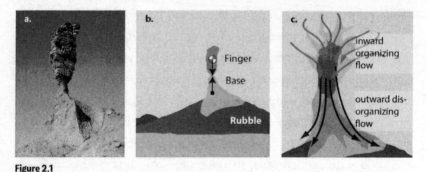

Figure 2.1
The Mukurob. *a*, As it stood before 1988. *b*, Equilibrium of forces in the static disequilibrium of the Mukurob. The Mukurob's center of mass is represented by the target. *c*, The Mukurob as a focus of dynamic disequilibrium.

which means "Bottle."* The Mukurob was a 34-meter-tall column of sandstone, supported on a thin pillar of softer mudstone below (Figure 2.1a). The softer rock had been relentlessly eroded by wind, leaving the Mukurob perched on an increasingly precarious pedestal, until the thing finally collapsed in 1988 (Figure 2.2). Before that, it had been one of the most photographed rock formations in the world.

The Mukurob's collapse was an economic blow to the impoverished Nama sheepherders who lived nearby, because it had provided a modest stream of revenue from tourists willing to part with a little cash for an enterprising guide, or a springbok skin rug thrown into the bargain. Once the Mukurob toppled, the trade in photography of unusual rock formations shifted north in Namibia to the Vingerklip ("Finger Rock"), which has been a boon to the more northern Damara tribes on whose land it sits. The Nama have not taken this

---

* There are different theories about the origin of the name. One legend points to a rivalry between the Nama and the cattle-herding Herero people, who were encroaching on the Nama from the north. The Herero, rich in fat cattle, mocked the Nama as having "nothing but rocks." In response, the Nama pointed to the Mukurob as a special rock that connected them to their gods and the land. They invited their Herero taunters to use as many of their fat cattle as they could to pull the rock down, which they failed to do. In response, the Nama shouted "Mû kho ro!," or, "There, you see!"

Figure 2.2
The Mukurob in 1998. The remnants of the mudstone pedestal still stand at the summit. The Mukurob itself at equilibrium consists of the rubble field strewn to the left down the slope.

lying down, though: by the peculiar logic of ecotourism, this has led to an effort to erect a fiberglass replica of the Finger of God, an idea that was shelved for a time as being a little on the crass side but that, as of this writing, was enjoying a comeback.*

I am telling you about the Mukurob because it offers some interesting object lessons in matters of stability, equilibrium, and disequilibrium—matters in which homeostasis is deeply embedded and which are easily and often confused. To clarify why homeostasis should be seen as something more than mere mechanism, we need to unpack those assumptions.

Let's start with the idea of stability, to which homeostasis is often likened. We describe something that is unchanging as "stable," so when we observe, say, a body temperature that is unchanging, we

---

* To no end yet, however. I visited the site in 2015, and the Mukurob is still a pile of rubble.

naturally speak of a "stable" body temperature. Stability, meanwhile, is often conflated with a state of equilibrium, balance, or repose. A "stable" body temperature is somehow an equilibrium body temperature. Therein lies the first trap.

All physical systems tend to go naturally from instability to stability, and it is the Second Law of Thermodynamics that tells us so. Stability and equilibrium thus do indeed go together, but this usually comes about when a system comes to rest at some minimum state of repose. This description applies to the rubble pile now strewn at the base of the erstwhile Mukurob's mudstone pedestal. However, this state does not apply to the Mukurob as it once stood. The Mukurob itself came to be at a stable equilibrium in 1988 when gravity, erosion, and a nudge from a temblor eventually brought it stably to rest as a rubble pile at the lowest position it is likely to occupy. In a living system, this kind of stability is not homeostasis; it is death.*

A somewhat more subtle idea is that homeostasis is a peculiar kind of *instability*. That's a better way to think about it, but instability comes in a variety of flavors, some of which qualify as homeostasis, and some of which do not. For example, it is clear in hindsight that the Mukurob was *un*stable. Before its collapse, the Mukurob seemed to be stable, but this state was maintained through a straightforward balance of just two static, but precariously opposed, forces. Gravity pulled the Mukurob down with a force directed vertically downward through its center of mass. The Mukurob nevertheless stayed steadily in place because its weight was met by an equal and opposing force pointing *upward* from the mudstone base (Figure 2.1b). As

---

* There is another interesting local legend about this rock, attributed to the San, that the Mukurob would stand as long as the white people ruled over Namibia. Once Namibia gained its independence in 1990, this legend, almost certainly apocryphal, was "recalled" as a typical example of the power of retrospective prophecy. It worked the other way, of course. Those of a dire frame of mind regarding Namibia's independence under the ruling party of SWAPO (South West Africa People's Organization) turned the prophecy around and regarded the fall of the Mukurob as a portent of impending disaster.

long as these two forces met head-on, the Mukurob stood—steadily *out of equilibrium.*

Now here is the crucial thing to understand. As long as the forces were aligned, the Mukurob's disequilibrium stood steadily with *no effort* being exerted to maintain it. It did, however, take effort to *undermine* the stability. The Mukurob collapsed following some perturbation, an earthquake or an errant gust of wind, which set the Finger of God wagging ever so slightly until its center of mass was pushed sufficiently out of line with its mudstone base so that it toppled over. So, we can say that before its collapse, the Mukurob was in a kind of *static dis*equilibrium, persisting as long as the precarious balance of forces was maintained but collapsing catastrophically when the balance was sufficiently perturbed.

This doesn't seem to fit homeostasis very well either: the most striking thing about life is how *robust* it is in the face of perturbation, compared with how frail the Mukurob's fragile static disequilibrium turned out to be.

Homeostasis is also a form of disequilibrium, but to appreciate how it differs from the Mukurob's static disequilibrium, we must let our imaginations run loose a bit. Now, imagine the Mukurob not as a solid mass of sandstone sitting on its narrow pedestal but comprising an ephemeral cloud of loose sand grains arranged in the Mukurob's distinctive shape. In this fantasy, there is nothing holding these sand grains together, and so gravity pulls them down in a sandy cascade raining down from the imaginary ephemeral Mukurob (the "outward disorganizing flow" in Figure 2.1c). Ordinarily, this ongoing loss of sand would make the Mukurob disappear, but—here is the strange part—imagine the Mukurob persisting, seemingly stable in its appearance, despite the cascade of sand falling from it. How can this happen?

There is only one way: the continual loss of sand must be offset by another stream of sand flowing *into* the imaginary Mukurob, just as fast as it is lost (the "inward organizing flow" in Figure 2.1c). Even stranger, for the imaginary Mukurob to retain its distinctive appearance, this cloud of sand streaming into the Mukurob must become organized in precisely the shape and orientation of the sand grains that have fallen away. Even stranger still, this persists even in the face of the high winds, pelting rains, or baking heat and shaking earth that ordinarily would (and ultimately did!) perturb the real formation's shape.

Our imaginary Mukurob is not so much a *thing*, therefore, as it is a *process*—a continual flow of sand that first becomes highly organized and then dissipates into disorder as it falls, all the time maintaining a persistent state of highly specified and complex organization. In short, our fantastic Mukurob is not in a state of *static* disequilibrium, as the real Mukurob was, but is actually in a state of *dynamic* disequilibrium.* And whereas the static disequilibrium of the real Mukurob did not require any effort to maintain, work has to be continually done to sustain our imaginary Mukurob: the sand might fall out on its own, but it takes effort to lift and precisely arrange those innumerable sand grains streaming into the Mukurob's particular form.

This is the essential nature of homeostasis: it is *life as a persistent dynamic disequilibrium*. Remember this phrase and let us reflect on its meaning, for we will be coming back to it again and again through this book. Our imaginary Mukurob was a fantastic realization of this idea. Yet that self-adjusting confluence of forces is precisely what living things do as a matter of routine. It is what distinguishes the living from the inanimate world, and that world's relentless slump into equilibrium and death. As I've said, homeostasis is an exceedingly strange idea.

This brings us to the dilemma posed by Bernard's dangerous idea. If we were wandering in the desert and came across our imaginary

---

* The Israeli chemist Addy Pross has designated this as "dynamic kinetic stability," or DKS, although it is far from stable.

Mukurob, the continually self-correcting rock of our fantastic imagination, we would probably assert that it is a trick and someone is secretly manipulating the sand. Or we may start believing it is governed by magical forces. The beauty of homeostasis as Bernard phrased it—as the constancy of the internal environment as a condition for life—offers us a way to avoid falling into that trap, but it means we have to think about life and its nature in some unaccustomed and perhaps uncomfortable ways.

———•———

Yet open any modern textbook on physiology, and you will not find any strangeness there. This is because the weird mystery of homeostasis has been captured, corralled, and tamed behind a philosophical fence of mechanism, concerned only with the "how" questions, not the "why."

Being a physiologist, I can hardly disagree with this: the elegant and complex workings of homeostasis *are* endlessly fascinating, something akin to how a mechanic must feel when he or she opens the hood of, say, a Lamborghini Murciélago.* Once the thrill passes, though, both the physiologist and the mechanic are charged with focusing on practical things. Wonder will not tell us how the Lamborghini works or how to fix it; mechanism will. Indeed, wonder can positively get in the way. Still, one cannot look at a beautiful

---

* The Murciélago is the latest incarnation of the so-called Lamborghini super cars, built around Lamborghini's LP640 engine, with 6.4-liter displacement, twelve cylinders, and numerous design innovations to effectively harness the engine's power. As is the tradition with the Lamborghini company, the name of the car derives from the sport of bullfighting. The car is the namesake of a legendary fighting bull from an 1879 match in Cordoba during which the bull kept fighting despite enduring more than two dozen strokes of the sword. So impressive was the bull's passion that the matador, Rafael Molina Sánchez, spared the bull's life, a rare honor. Murciélago himself went on to found the Miura, one of the most famous of the lineages of Spanish fighting bulls (though some doubt has been cast on this story).

creation like the Murciélago and not see in it a kind of beautiful soul. Ah, we sigh as we awaken from our reverie and snap ourselves back to cool rationality, which is the talk of the romantic, not the scientist or mechanic. To talk of souls is to tread into the realm of the religious, of spirits, spooks, and goblins defining life, areas where the scientist or the mechanic is not allowed to go.

Yet the nagging doubt remains. There is an awful* beauty in the Lamborghini that makes it somehow *more* than a machine, something infinitely richer than a mere machine. The same, exponentially squared, can be said of life's awful beauty, and this was Bernard's whole point. A rich understanding of life somehow has to account for this awful beauty. Where does it come from? What does it mean? More to the point, what if a relentless focus on mechanism leads us to a crabbed and impoverished view of what a Lamborghini is? Wouldn't we have a richer understanding if we could know the *soul* of the Murciélago—even better, to know its soul, not just as a romantic, but as a mechanic?

This is the dangerous dilemma that homeostasis poses to the modern science of life. We are called to focus only on the cogs and gears, but when we do so, we strip away a soul. We shove into the back of our minds something fundamental and enriching about the "why" of homeostasis, something that will give us the key to understanding its abiding strangeness. The dilemma is made all the more acute when we look back and see what Bernard was trying to convey: he was not asking us to admire cogs and gears, he was inviting us to contemplate a beautiful soul.

In the Preface, I promised you strangeness, and we have indeed wandered into some very strange territory: of fantastic rocks, soulful automobiles, and wonderful life. I have led you here for a purpose, though, because abiding strangeness has permeated our thoughts

---

* In the archaic sense of the word: inspiring reverential wonder.

about life for millennia. We do not teach students much about this strangeness. Rather, we dismiss it as an antiquated distraction from the real business of becoming scientists. We ignore this history at our peril, though, because the strangeness of the living organism was at the heart of Bernard's dangerous idea.

3

MANY LITTLE LIVES

It is important that we get to know Claude Bernard and his thought a little better. His fame rests upon his founding of the science of *experimental* physiology, which is reflected in the title of his 1865 work *Introduction to the Study of Experimental Medicine*. The book is, in its own way, as foundational a work for medicine as Darwin's 1859 *On the Origin of Species* was for evolution, and for similar reasons. Just as Darwin supposedly banished the ineffable from the phenomenon of evolution (a claim that we shall see is not quite true), so too did Bernard supposedly banish the ineffable soul from the phenomenon of life. Armed with a sharp mind, and sharper surgical tools, Bernard is renowned as the rationalist warrior who banished forever the ghosts and vital spirits that had, until then, haunted physiology and medicine.[1] After Bernard, we would never go back to that world. Just to underscore the triumph, we have a scarlet letter we can pin on those tempted to backslide. It is a scarlet V, for *vitalism*, the V-word, to be prominently worn as a warning to anyone who flirts with the idea to venture there, lest they, like Nathaniel Hawthorne's Hester Prynne, be shamed and shunned.*

---

* In Nathaniel Hawthorne's 1850 novel, *The Scarlet Letter*, Hester Prynne was forced to wear a scarlet letter "A" on her clothing to signify her adulterous relationship with the minister of her church, Arthur Dimmesdale.

But hold on. In the last chapter, didn't I just write that Bernard was asking us to contemplate the beautiful soul of the organism? How does that square with the conventional narrative of Claude Bernard as the killer of vitalism? How can beautiful souls even have a place in such a mind as Bernard's?

But beautiful souls do belong, and here is why. If we shift the light slightly on Bernard's hagiographic portrait, some interesting wrinkles pop out. During Bernard's lifetime, for example, what we now regard as his signature idea—homeostasis—had almost no effect and was almost never mentioned or discussed by his contemporaneous colleagues. Nor was it mentioned by his eulogists.[2] Indeed, physiologists didn't really pick up on the idea until some seventy years after Bernard published his famous aphorism. When homeostasis did reemerge, early in the twentieth century, it was in a cramped and desiccated form, detached from Bernard's original intent (more on that in the next chapter). Even today, there seems to be an enduring ambivalence about Bernard's signature idea:[3] a memoir of him published as recently as 1989 does not even mention the word "homeostasis" or refer to the idea.[4] It's almost as if we're ever so slightly—embarrassed? And perhaps there's reason to be, because Bernard, far from being the scourge of the V-word, was actually its vindicator.

The roots of this contrarian conclusion are to be found in vitalist thought as it was transformed through the eighteenth and early nineteenth centuries.[5] This is a history we generally do not teach to students today, in part because of the bad odor that presently clings to vitalism; but it also stems from the imagined separateness of science from all other intellectual endeavors. While historians and philosophers might find the history of vitalism a fit subject, we scientists might say this attention to finding the secret ingredient for explaining life would only be a distraction for those interested in "real" biology, which must be impeccably mechanistic and divorced from philosophy.[6]

It is also not taught, it must be said, because it is inconvenient to the narrative with which modern biology prinks itself—as a science that has painstakingly climbed up from the vitalist darkness and into the rationalist light. The conventional narrative about Bernard fits this pretense very nicely, which, as these things often go, elevates the simple narrative above the complicated truth. And it is inconvenient for the same reasons a distinguished family might try to cover up the horse-thieving ancestor who established the family fortune: most of modern biology actually finds its roots in the supposedly disreputable vitalist thinking of the nineteenth century, including—most inconveniently of all—Darwin's evolutionary thought.

The modern sciences of life cannot be properly understood without an appreciation of this vitalist core. Shoving it into the shadows, as we commonly do today, therefore amounts essentially to a form of narcissistic preening; we are blinded to the world around us in favor of the pretty baubles of mechanism we have adorned ourselves with. Don't take my word for it: try mentioning vitalism to a room of biologists today and you will usually be met with polite indifference, or a troubled darting of eyes.

—•—

Vitalism poses two basic questions. We already chewed these over in the Preface, but they are worth repeating: is life a special phenomenon, unlike any other in the universe, and if so, what makes it so? If your answer to the first question is "clearly yes," you are a vitalist, whether you want to admit it or not. The rub comes with how the second question—what makes it so?—is answered.

Traditionally, that answer took the form of what we may call *essentialism:* life is special because it is imbued with a special vital essence, or *vis essentialis*. Life exists because this *vis essentialis* infuses and animates otherwise inanimate matter. This notion is venerable to the point of being hoary. It predates the Hippocratic physicians,

whose medical philosophy derived from the still more ancient doctrine of *humors:* systems of opposing forces, like the lightness of air
versus the heaviness of earth, or the heat of fire versus the cold of
water. Life, in this view, was a state of perfect harmony suspended in
a matrix of diametrically opposed forces, with the balance mediated
by the *vis essentialis*. To the Hippocratic physicians, health was a state
of balance and harmony. Sickness, disease, and death were states of
imbalance and disharmony.[7]

This doctrine was the theoretical core of medical practice well
into the eighteenth century. Therapeutic bloodletting ("bleeding" a
patient) provides an interesting example of this philosophy translated
into practice.* Traditionally, therapeutic bloodletting was justified by
the need to release from the patient an excess of one vital humor that
was out of balance with another. Release the excess humor, and the
balance of humors would be restored, as would be the patient to a
state of health. The practice, indeed so much of medical practice in
those times, probably killed more patients than it helped, but never
mind, it was justified by sound and venerable teaching—the science
was settled, we might say today.[†8]

By the eighteenth century, vitalism, and the medical practices it
engendered, had become the subject of vigorous debate between

---

* In therapeutic bloodletting, sometimes as much as half the patient's blood volume was removed from
  circulation. The amount of blood to be drained for the therapy to be deemed effective was usually
  determined by the point at which the patient fainted.

† Therapeutic bloodletting survived into the early twentieth century and provides a remarkable object
  lesson in the cultural inertia that imbues medical practice. Bloodletting had long been justified as a
  means of draining an excess of various humors from the body, restoring balance. Well into the nineteenth century, by the time the notion of humors had fallen into disfavor, bloodletting continued to
  be the therapy of choice for certain types of fever, notably "sthenic" fevers, which were marked by
  agitation and hyperactivity, and was justified by its "calming" effect on the patient. Bloodletting also
  persisted as a treatment for "dropsy," known today as congestive heart failure. In the early twentieth
  century, bloodletting fell into near-complete disfavor and was denounced as quackery. The practice
  continues, however, as the therapy of choice for certain disorders of excess blood cell production
  (polycythemia and hemochromatosis). Bloodletting of sorts through the use of medicinal leeches is
  returning as an accepted medical practice, mostly for treatment of necrosis and hematoma.

competing European schools of academic medicine.[9] What emerged from this debate was a radical transformation of vitalist thought from an "essentialist" vitalism (sometimes called "metaphysical vitalism") to a "process" vitalism (also called "physical vitalism" or "scientific vitalism").[10] Through this subtle shift in perspective, vitalist thought came to be more concerned with action and mechanism than with some ineffable "vital stuff." In this way, the seeds were planted for metaphysical vitalism's eventual rout in the nineteenth century.

Initially, the newly emerging process vitalism drew deeply from its essentialist roots, taking as its inspiration another vital essence, the *vis mediatrix,* or mediating essence. This vital substance was cooked up as a means of explaining how the actions of the body's various parts could be coordinated. It was thought to be dispersed through the nerves and would link, or mediate the interactions between, one body part, say the heart, and another, say the stomach. Despite its theoretical utility, *vis mediatrix* proved to be the poison pill for essentialist vitalism. In keeping with the rising tide of Enlightenment skepticism and the willingness to resort to the authority of empirical proof over that of venerable tradition, the hypothesis of the *vis mediatrix* opened up a process-oriented outlook toward medicine that proved more congenial to posing experimental questions. The whole doctrine of vital essences could not stand up long under such scrutiny. *Vis essentialis* was the first to go: physicians could offer only the *fact* of the living body as evidence of its existence, or a corpse as evidence of its absence. In short, *vis essentialis* died because it was a scientific dead end.

That did not entirely kill the idea of vital essences, though. Once *vis essentialis* was out the door, new essences were cooked up to replace it, because the scientific consensus was that there simply *had* to be vital essences.[11] The *vis mediatrix* was one of many designated hitters brought in to fill the roster. But these couldn't stand long either, for the same reasons that did in *vis essentialis:* they were slippery ad hoc concepts rather than real things you could hold in your hand or collect in bottles.

Nevertheless, a new spirit of skeptical and evidence-based scrutiny of nature was wafting through European medicine. Much of this nascent empiricism was anatomical and observational. One can find in writings from those times catalogues of correlations between certain diseases and modifications of the body's organs, for example. All well and good, but there was little in the way of a radical challenge to the prevalent essentialist thinking. Some of those observations, though, were experiments that called into question the very existence of the mediating "stuff." Why, for example, could an organ, like a frog's heart, survive for a time outside the body, removed from the influence of any supposed *vis mediatrix*, and still be responsive to changes in its environment? Why, if the nerves were the conduits for the *vis mediatrix*, could some organs, like the pancreas, function even though the organ had no discernible innervation?* What is important here is the empirical and skeptical mindset of these early process vitalists, in which contrarian phenomena, such as the disembodied frog heart, were not simply anomalies to be explained away by cooking up another vital essence, but were disproof of a long-standing doctrine.

Among the most interesting of these contrary observations came from Théophile de Bordeu (Figure 3.1) of the medical college at Montpellier in southern France. Montpellier was one of three medical colleges that drove the transformation of vitalist thought in the eighteenth and nineteenth centuries.† In a remarkable eighteenth-century anticipation of the superorganism idea,‡ Bordeu drew a

---

* We now know these phenomena are mediated in part by circulating hormones, which, ironically, are coming to look a lot like the chemical neurotransmitters that nerves employ to mediate their control of the body. Maybe there are many *vis mediatrices*!

† The two others were Leiden, in the Netherlands, and Halle, in Germany.

‡ The "superorganism" is often invoked as a metaphor for the colonies of social insects like bees, ants, wasps, and termites. See more in Chapter 9.

Figure 3.1
Théophile de Bordeu (1722-1776).

parallel between the coordinated behavior of a swarm of bees and a living body. If bees in a swarm were separate bodies, physically disconnected from one another, how could any putative *vis mediatrix* flow between them? And if there was no *vis mediatrix*, how were the behaviors of the individuals mediated to produce the hive's "organism-like" behavior? Bordeu's little thought experiment not only called into question the very idea of a *vis mediatrix*, it opened the door to a new and powerful metaphor for understanding life, its function and dysfunction—and as we shall see, its evolution. This was the metaphor of life as an assemblage of "many little lives," of organisms as collections of semiautonomous units that, through a process of negotiation and mutual accommodation, produced the coherent organism. This was a new way to think about the vital organism. It was not the "stuff"—the vital essence—that formed the organism, but the *process*—the negotiation and mutual accommodation of the organism's "many little lives."

Thus was the stage set in 1834, when Claude Bernard went to Paris as a young man, eventually to be drawn into the orbit of the remarkable François Magendie (Figure 3.2). Magendie is important to our story for a number of reasons. He was a caustic critic of essentialist vitalism; he was a flamboyant empiricist and rationalist of the highest order; he became Bernard's mentor and collaborator; and he plowed the ground that Bernard later sowed so fruitfully.[12]

François was the scion of a prickly revolutionary family. His father, Antoine, was a "barber-surgeon," practicing at a time when physicians held that profession in some disrepute. As is often the case with upstart professionals fighting for legitimacy and respect, Antoine was aggressively antiaristocratic, anticlerical, skeptical, and rational. When the revolution came, Antoine relocated his family, including the nine-year-old François, to Paris, so he could be in the thick of the upheaval. Eventually, Antoine became disillusioned with practicing surgery (not surprising in light of the horrific

Figure 3.2
François Magendie (*left*, 1783-1855) and Claude Bernard (*right*, 1813-1878).

suffering inflicted on surgical patients in those preanesthetic days).*
So he quit his profession to serve the new republic, which proved
to be an impecunious decision. Despite his disillusion with his erstwhile career, Antoine determined that his son should not follow in
his father's folly and instead should seek a career in medicine as his
father had. Unfortunately, Antoine had raised his children in such
a way that the young François was woefully unprepared for admis-
sion to the prestigious École de Santé where his father had wanted
him to go.† Antoine's connections among his former physician colleagues came to the rescue, though, and François was taken on as
an apprentice to Alexis Boyer, a prominent French barber-surgeon
who, like Antoine, had also risen from humble roots. Under Boyer's
supervision, François flourished and within a year had passed the
examinations qualifying him to become a hospital intern. Four
years after that, he entered the medical school of the Hôpital Saint-
Louis. There, he distinguished himself as a skilled anatomist with a
lively—some said prickly—mind. In addition to studying medicine,
Magendie was asked to give courses in anatomy and physiology at
the École de Médecine. Once he received his medical degree in 1808,
Magendie's path to a stellar career in medicine seemed to be set.

What followed was an erratic career of brilliant success interspersed with controversy and dissipation. Although Magendie
quickly had realized his father's aspirations that he become a physician, and had established himself as a teacher of exceptional skill

---

* The horrors of surgery and medical practice prior to the introduction of ether anesthesia in the 1840s
are vividly and mordantly described by Dr. Lindsey Fitzharris in her blog *The Chirurgeon's Apprentice*
and in her 2017 book *The Butchering Art: Joseph Lister's Quest to Transform the Grisly World of Victorian
Medicine*.

† Antoine Magendie raised his son according to the principles spelled out by Jean-Jacques Rousseau in
Emile, or Treatise on Education (1762), which argued that education of children is best accomplished by
allowing them to flower on their own rather than be tamed by the rigor and discipline of the traditional classroom. As a result, François did not learn to read or write until he was ten years old. He was
not stupid, however; he demanded at the age of ten to be admitted to a regular school and progressed
rapidly to winning national essay contests by the age of fourteen.

and accomplishment, he had difficulty sustaining a medical practice because he was denied an appointment as a hospital physician and blocked by senior colleagues who regarded him as a dangerous rival. Magendie was not his own best friend here: he fell into acrimonious disputes with colleagues within France and across Europe and the British Isles. He became the beneficiary of an inheritance only to squander it on horses, parties, and drink. For a time, he offered public dissections and vivisections for a fee. His demonstrations commanded high prices, because of Magendie's extraordinary surgical skills, his showmanship, and his willingness to make himself the subject of some of his demonstrations. Being shut out of medical practice, he turned his considerable energy and curiosity to medical research, bouncing from one appointment to another in the Paris medical scene. He made remarkable discoveries in the function of the brain. Ultimately, even his enemies could not hold this brash man back: he was eventually elected to the Académie des Sciences and to the Académie Royale de Médecine ("in spite of myself," in Magendie's own ironic words).

Magendie's erratic, critical, and abrasive personality and his energy, bluntness, and irritation with received wisdom grabbed French medicine by the collar. The particular target of his scorn was the essentialist idea, and at the time of Magendie's rising fame, it was starting to gasp its last, propelled there in no small part by Magendie's pointed attacks on the endless proliferation of "vital forces" that academic physicians were wont to invoke whenever an unexpected phenomenon cropped up:

> Why then is it necessary in respect to every phenomenon of the living body to invent a peculiar and special vital force? Cannot one be content with a single force which one could designate "vital force" in a general way, while admitting that it gives rise to different phenomena depending upon

the structure of the organs and tissues which function under its influence? But is not this single vital force still too much? Is it not an hypothesis pure and simple, inasmuch as we are unable to perceive it?

Magendie was more than capable of delivering this criticism in the bluntest of terms:

> To express an opinion [on the existence of a vital force], to believe, is nothing else than to be ignorant. . . . One could with justice say to you "You believe, therefore you don't know."[13]

One can easily imagine the attraction the brash Magendie must have had for the young Bernard, in from the provinces and looking to make his mark.

———— • ————

Bernard's upbringing was more conservative than Magendie's. His father, Pierre-François, was a provincial winemaker and *petit* land-owner in Beaujolais, and so he was not as caught up in revolutionary fervor as was Antoine Magendie. Claude Bernard's early education was mostly in local Jesuit schools. These bored him immensely, but he had an active and curious mind, so he became an autodidact, indulging a growing interest in philosophy and the romantic arts. One had to make a living, though, so when a school friend enthused about his new career in pharmacy, Claude decided to follow his friend and become a pharmacist's apprentice himself. This proved not to be as happy an experience as he had hoped, so Claude surreptitiously began to plot his escape to Paris, to follow there his dream of becoming a playwright. The plot ripened prematurely when, in the aftermath of an accident at the apothecary, Claude's pharmacist master discovered that Claude had been working on a play rather than attending to potions. He was duly shown the door. Seizing the opportunity, off to Paris he went.

Once there, Bernard's artistic aspirations quickly foundered at the hands of the critic Saint-Marc Girardin, who kindly, but firmly, advised Bernard that he should perhaps try his hand at medicine instead. Wisely, Bernard took this advice (thank you, Girardin!), and after some difficulty with the entrance examinations, he enrolled in medical school. To support himself, Bernard took on work as a laboratory assistant in a Parisian girls' school, where he proved to be extraordinarily skillful in dissection. Bernard's anatomical prowess soon captured the attention of Magendie, who took him on as *preparateur* for his lectures at the Collège de France. Soon thereafter, Bernard formally became Magendie's student, setting him firmly on his own path to a scientific career.

Bernard's relationship with the mercurial Magendie was both contrary and complementary. It was stormy, marked by intermittent fallings-out and reconciliations. Like Magendie, Bernard brought to his work a healthy dose of skepticism, seeded in this case by his recollection of the slapdash pharmaceutical formulations (which sometimes included the leavings of other potions) that he had been ordered to make during his apprenticeship, and by the utter lack of rationale or evidence for their effectiveness. Bernard's career was as mercurial as Magendie's. Like Magendie, he had difficulty as a practicing physician, although unlike Magendie, Bernard never felt compelled to practice medicine, preferring research. He left Paris to return to his home village to establish a country practice there, but his heart wasn't in it. His marriage to the wealthy Fanny Martin gave him the freedom to pursue his research interests, but the freedom was bought at a terrific price of an unhappy marriage and the premature deaths of all three of their children. Like Magendie, his brilliance as an experimentalist and teacher propelled his eventual rise to the top of French science. He was made a Chevalier of the Légion d'Honneur; he obtained a doctorate in zoology to bolster his research credentials; and he was eventually appointed to a special chair at the Faculty of Sciences in Paris, was elected to the Académie des Sciences and the Académie de Médecine, and took over Magendie's chair of medicine after Magendie died.

Bernard was attracted to Magendie's thoroughgoing skepticism and his willingness to challenge conventional wisdom. Bernard's skepticism was rooted in more conservative soil than Magendie's, though. While Magendie's radical skepticism bolstered Bernard's own, Bernard parted company with his mentor in some significant ways, the most significant being over the nature of life. Magendie was a radical materialist, prepared to fold life fully into the nascent field of chemistry. Bernard was not quite prepared to go so far, though, which is captured in his aphorism: the stability of the *milieu intérieur* is the *antecedent to,* not the *outcome of,* the free and independent life. An understanding of the chemical workings of life would enable physiologists to meet their prime scientific obligation: to make the workings of life's antecedent intelligible. But physiologists could never forget that it is the antecedent, life's unique nature, that is life's principal "fact on the ground." Bernard expressed it this way:

> Since physicists and chemists cannot take their stand outside the universe, they study bodies and phenomena in themselves and separately without necessarily having to connect them with nature as a whole. But physiologists, finding themselves, on the contrary, outside the animal organism which they see as a whole, *must take account of the harmony of this whole,* even while trying to get inside, so as to understand the mechanism of its every part. The result is that physicists and chemists can reject all idea of final causes for the facts that they observe; while *physiologists are inclined*

to acknowledge an harmonious and pre-established unity in an organized body, all of whose partial actions are interdependent and mutually generative [emphasis mine].[14]

Here we see the elements of nineteenth-century process vitalism—the notion of the organism as a harmonious whole comprising "many little lives," along with the emphasis on teasing out the processes of mutual accommodation that undergirded the principal "fact on the ground": the integrated, coherent, and harmonious organism.

Bernard's fame, therefore, rests on two foundations. The first is the one we celebrate today: he is the brilliant experimentalist who teased out the mechanisms of living things. This legacy is what underpins our modern conception of life as a mechanism. The other comprised the more philosophical speculations that grounded Bernard's thinking about the nature of life. Today we greet these with, at best, a perfunctory nod. Our modern diffidence streams from Bernard's inconvenient vitalism and his solution to vitalism's fundamental question: if life is a unique phenomenon of nature, what is it that makes it distinct? To Bernard, life was irreducibly unique, and what set it apart was homeostasis. This is where we encounter the modern misconception about homeostasis: it is not a statement of rational mechanism; rather, it is a profoundly vitalist idea.

Bernard's conception of homeostasis traces its roots back to the previous century's fertile turmoil over essentialist versus process vitalism. Among the eighteenth-century vitalists was a growing difference of opinion on what an organism's "many little lives" were. Some thought they were the innumerable (and recently discovered) cells of the body. Some thought they were the organs of the body. And some, like Bordeu, saw the "many little lives" existing at multiple scales, even among assemblages of organisms like bee colonies. Despite this, there was a broad consensus among the process vitalists that life's distinct quality emerged from the negotiation and accommodation of the organism's innumerable "little lives" to one another.

One of the outward signs—what we might today call an "emergent property"—of this ongoing mutual accommodation was adaptation. Organisms can exist in a variety of circumstances and respond in predictable and repeatable ways to ensure their continuing "good fit" to whatever environment they are in: thicker fur in the winter, increased strength when worked, and so forth. This is the meaning of adaptation: a tendency of living things to "apt function," to have an aptitude, if you will, to persist in the face of a whimsical environment.

Bernard's novel take on the "many little lives" metaphor was to internalize the phenomenon of adaptation. The persistence of a life in the face of environmental perturbation had to involve a certain independence from the environment: Bernard's "free and independent life." Thus, body temperature would be steady even as outside temperature varied, intervals of starvation and thirst would be tolerated, and so forth. This independence could be disrupted experimentally. Cut a nerve here, occlude a blood vessel there, remove this or that organ, and the innumerable internal conversations between the organism's "many little lives" would be disrupted and their striving to mutual accommodation thwarted: body temperature would fall, the heart would race, the stomach would cease to churn.

To Bernard, there had to therefore be a process of mutual accommodation operating within the body that mirrored the adaptation of the body itself to the fickle and variable external environment. Just as the body was immersed in an external environment, so too were the body's "many little lives" immersed in an *internal* environment, namely, the medium of the body's internal fluids. And if adaptation to the external environment sustained the organism, perhaps the adaptation of the body's "many little lives" to one another, whatever those little lives might be, produced a sustained internal environment. Bernard sought to untangle all this mutual accommodation through the study of this internal environment, the *milieu intérieur* of his aphorism, which he did brilliantly in his experimental work. But hovering over it all was the core of nineteenth-century process vitalism: life as the mutual accommodation of "many little lives."

Bernard's skill at dissection and experimental design made him the master of asking and getting credible answers to the "how" questions of the "many little lives," and his tenacity at running the questions to the ground until they yielded an answer is why he is justly remembered as the founder of experimental medicine. In this sense, Bernard was indeed the nemesis of vitalism, but only in the limited sense of the dwindling metaphysical variety: once he was done with it, that form of vitalism would never come back. Yet Bernard's approach to experimental medicine cannot be understood independently of the larger motivation behind his work: to vindicate a profound truth about life, that life is a special *quality,* that chemistry might be a tool for studying life but it is not life itself. It is matter that serves life and not the other way around. In short, Bernard's work was actually a strong empirical defense of the *process* vitalist thought of his day.[15] Bernard was not vitalism's nemesis—far from it—but its vindicator.

Let us now step back a bit and consider where we have been, because it will help make sense of what is to follow. Bernard's story is that of the double-edged sword of the Hobson's choice that sits at the heart of modern biology, which we may now rephrase: can one be a scientist and a vitalist at the same time? Modern biology's answer to this is "no." To be a scientist who studies life, you must also be a mechanist and a materialist. You emphatically may not be a vitalist. To do so is to don the scarlet V.

Claude Bernard stands as a striking refutation of this assertion. This is inconvenient to the modern, and it must be said, essentially philosophical stance of materialism as the only credible way to think scientifically about life. When we are confronted with someone like Bernard, two things commonly happen. We might find, for example, that a clean narrative begins to assume precedence over the complicated truth. If Bernard was a brilliant practitioner of the experimental arts, then his vitalism becomes simply inconceivable and cannot be part of the narrative. This is an essentially political dynamic—a dynamic we will encounter again and again throughout this book—because narratives are enforced through political means. There is another motivation at work, this one at least having the virtue of intellectual credibility. To wit: is there a way to reframe what is essentially a vitalist idea into credible mechanism? In short, is there a way to render vitalism irrelevant? Can we explain away life's unique nature as illusory? We also will encounter this dynamic throughout this book, starting with what comes next: the revival of Bernard's vitalist conception of homeostasis in the triumphant materialism of the mid-twentieth century.

4

A CLOCKWORK HOMEOSTASIS

If Claude Bernard was vitalism's vindicator, then why did vitalism become the V-word? It's not as if Bernard or others like him were proposing a science of angels and demons. What explains the opprobrium? Answering that question could fill volumes, but the basic motivator was a tectonic shift in the metaphysical universe that began in the latter years of the nineteenth century and built for several decades into the twentieth. In the end, our modern attitudes toward the value and reach of science came to be cemented into place, setting modern biology on its trajectory toward its impending crisis.

The shift was driven by a tension that has riven Western philosophy ever since Socrates. It turns on a ten-dollar word: *epistemology*, or how we come to know and understand the world. How we proceed on that quest turns on whether we think the world is all mechanism, including the life that inhabits it, or whether a broader organizing principle makes it all what it is.[1] If we think it is the first—all mechanism—then understanding comes from drilling down to the uttermost tiny details: reductionism, materialism, mechanism, in other words. Transcendence is called for in the second case, the willy-nilly details of mechanism being just that—willy-nilly details. Understanding from that perspective comes through identifying principles that transcend matter and mechanism and impose order upon it. The organizing principle could be God, the demiurge, vital stuff, or intelligence of some sort. Whatever it is, it will be *metaphys-ical*, that is, beyond the mere physical.

On the question of life, these opposed viewpoints swept like tides through the eighteenth and nineteenth centuries, trending one way, then the other, from the rational to the romantic then back, from the vitalist to the mechanist then back, endlessly surging but never seeming to settle on a "right" answer.[2] Bernard's form of vitalism pointed a way out of this seemingly Sisyphean state—a middle path that was not quite mechanism and not quite vital essence, but a hybrid, and an extraordinarily fruitful hybrid at that. But in the end, biology in the twentieth century did not follow Bernard's lead, choosing instead to simply try to stay the tide, never mind the surg-ing. There's a ten-dollar phrase for this as well—*epistemic closure*—where everyone simply agrees that we will think only one way about the universe—one form of epistemology—and not bother to engage other ways.

Epistemic closure has a rather bad odor about it, because it implies closed-mindedness—the phrase has lately been thrown around to smear one's political opponents as close-minded ignoramuses[3]—but closed-mindedness is not actually its problem. Indeed, epistemic clo-sure can be a form of mental discipline, a kind of Tao that, if followed scrupulously, can be immensely satisfying and rewarding. The rewards are aesthetic (seductive beauty revealed) and intellectual (captivating internal logical consistency), and chasing those rewards can engender extraordinary intellectual energy and creativity. This is why the twentieth century was a golden age for biology, even as biology became epistemically closed, that is, became mechanistic and reductionist to the exclusion of other ways of thinking about it.

There *is* a problem with epistemic closure, though: it can become a form of intellectual narcissism. Those living within an epistemically closed world become so engrossed in their beautiful, self-referential, and internally consistent universe that mental discipline easily lapses into mental blindness, followed by intellectual pathology and ultimate downfall. Among those pathologies is fractiousness, with intellectual energy directed to ever smaller problems or questions: you start by admiring the face, then focus on the slant and color of the eyes, then move to the intricate patterns of iridescent colors of the iris, and end up enthralled with the exquisite curvature of a single eyelash. In the sciences, we prefer to call this "specialization" rather than narcissism, but the dynamic is the same no matter what we call it: highly learned people coming to know more and more about less and less, until they know everything about nothing at all.[*] We see this trend rampant in biology today: there might be something we call "biology" that we teach to students; there might be academic departments of "biology" in our colleges and universities; but there are, in fact, dozens of "biologies" out there, with new ones coming along annually.[†4] Faced with this proliferation, it is reasonable to ask: is there such a thing as a coherent science of life anymore?

Another common pathology of epistemic closure is an increasing reliance on politically enforced orthodoxy. Narcissism demands that everyone admire the same thing, with unpleasant consequences for those who demur. The means to enforce orthodoxy can range

---

[*] According the *Yale Book of Quotations*, the aphorism, which is used frequently, probably traces its origin to William J. Mayo in the November 1927 edition of *Reader's Digest*: "A specialist is a man who knows more and more about less and less." See http://freakonomics.com/2009/07/16/quotes-uncovered-the-punchline-please/.

[†] One can judge the fracturing of a discipline by the numbers of scientific journals in which biologists publish. On the order of four thousand scholarly journals devoted to the life sciences are published today, and these have been growing at a rate of 3 to 4 percent per annum. See Ware and Mabe, "The STM Report: An Overview of Scientific and Scholarly Journal Publishing," (The Hague: The Netherlands, 2012).

from credentialing systems that ensure that only "right" thinkers are allowed to think professionally (which describes the modern university system), to sustaining agreed-upon myths by ridicule or expulsion of those who depart from the "correct" thinking, to exercising naked political power over dissidents through governments and courts of law.[5] The end-point, as it was for Narcissus himself, is isolation and alienation—of specialist from specialist, of scientist from artist or theologian, of scholar from public, of the science of life from the phenomenon of life itself.

Fortunately, narcissism normally is self-limiting: we snap out of it because reality tends to breach even the most assiduously built epistemic wall. What made the epistemic closure of the twentieth century distinctive was its reach and staying power. Once Friedrich Nietzsche declared in 1882 that God was dead, something had to fill the gap, and scientists came rushing in. This was the rise of scientism—science now as God—with claims to be the sole legitimate basis for thinking about all aspects of the world.[6] What followed was the predictable rise of "scientific" theories of history, economics, sociology, education, and politics. This was the era of the technocrats and the progressives, of rule by the credentialed over the incompetent uncredentialed, all inspired by the glittering phantasm of politics, economics, and even morality that could be made scientific, sensible, predictable, and rational—and therefore controllable.[7] We have never entirely snapped out of this latest venture into intellectual narcissism, an adventure that has endured for at least a century now.

Debate continues about whether we are better off or worse off as a result. We will not resolve that debate here. Suffice it to say that chasing that utopian promise jostled many long-standing verities off the stage—some that deserved to go, some less deserving of this fate—and an honest understanding of that time demands we sort out which were which. I am arguing that among the undeserving casualties was the novel form of nineteenth-century vitalism championed by Bernard and many others. The story of how Bernard's fundamentally vitalist conception of homeostasis became transformed into its modern anodyne, tamed, and neutered form of mechanism—a clockwork homeostasis, if you will*—illustrates the most pernicious feature of epistemic closure: its ever-increasing reliance on narrative, rather than evidence, to sustain it. This narrative can take many interesting and creative turns, as we shall see, but ultimately these run out, and a crisis follows. That is where we are today with respect to Bernard's "dangerous idea" of homeostasis and its implications for the "why" of life.

———•———

In biology, the new century's intellectual fashion was made manifest in what Ernst Mayr has called "physics envy"—the idea that, to be a "proper" science, biology had to make a decisive break from its vitalist past and biologists had to look to physicists and chemists for proper role models of how to do "real" science.[8] And that, in a very compact nutshell, is how biology came to be transformed nearly completely into the materialist and mechanistic discipline that is familiar to us today. It is also how the innocuous claims of the nineteenth-century scientific vitalists, Bernard included, came to be outré.

In physiology, where Bernard's conception of homeostasis should have found its safest refuge, physics envy prevailed as physiologists came to focus on the captivating details of mechanism being uncovered by new techniques in chemistry and microscopy. Physiologists—Bernard's heirs—thus turned Bernard's aphorism on its head. Life's chemistry was no longer a tool to probe life's special quality, as Bernard had done. Now, the chemistry of life became the

---

* With an obvious hat tip to Anthony Burgess's 1962 dystopian masterpiece *A Clockwork Orange*.

end in itself, with physiology morphing into "physiological chemistry," as this new approach came to be called.* The name sums up the not-so-subtle shift of attitude: we are chemists, "real" scientists, not stamp collectors, to echo the physicist Ernest Rutherford's infamous put-down of biologists and other nonphysicists.† We are not troubled anymore with vitalist obscurantism; life is chemistry, nothing special about it. Quasi-philosophical ideas, like Bernard's musings about the *milieu intérieur,* are distractions in the way of doing "real" science, mere philosophical decoration to what had become, in the fashion of the day, the repudiation of philosophy.

Despite the physiological chemists' efforts to get us all to look away, that nagging problem of the "many little lives" that had obsessed the process vitalists, including Bernard, kept popping up. In the early twentieth century the American physiologist Lawrence Joseph Henderson, for example, showed how regulation of the blood's acidity‡ involves a cooperative interaction between nearly every organ and cell type in the body; the "many little lives" phenomenon was thus embodied and personified, impossible to capture in glass culture tubes but glaringly evident in the arena of the whole organism.[9] Henderson's contemporary, Walter B. Cannon, saw the same propensity in the regulation of blood sugar, stress, and adaptive response to a changing environment. Indeed, he found this phenomenon so compelling that he did what Bernard, in his Gallic loquaciousness, had never done: he coined a single elegant word to

---

* "Physiological chemistry" was the precursor of our modern discipline of biochemistry.

† Ernest Rutherford (1871–1937) was a preeminent physicist of the early twentieth century who discovered, among other things, the nucleus and structure of the atom. He was noted for his pithy commentary on physics, chronicled vividly in J. B. Birks's 1963 memoir, *Rutherford at Manchester* (W.A. Benjamin). The quote cited in Birks was "All science is either physics or stamp collecting." Wikiquotes offers various unsourced variants, including my favorite "Physics is the only real science. The rest are just stamp collecting."

‡ Blood acidity needs to be robustly regulated, because small changes in the blood's acid balance can cause serious, even life-threatening disruptions in the function of entire organisms.

capture it—"homeostasis."[*][10] In the end, though, both Henderson and Cannon cast their lots with the surging materialist tide, leaving the more philosophical side of Bernard's legacy to drift in the twentieth century's intellectual backwaters. It would pop up here and there as eructations of the old vital essence—*élan vital,* entelechy, omega points, and so forth—and would inspire fringe schools of medicine such as homeopathy, which still lionizes Bernard as the inadvertent visionary founder of its curious philosophy of medicine.[11]

What sealed biology's materialist fate, though, was not physiological chemistry, but the rediscovery at the beginning of the twentieth century of the "atoms of heredity" of Gregor Mendel (1822–1884).[†] These had languished in obscurity after their discovery nearly sixty years previously, but once rediscovered, the big prize in biology became the hunt for the material nature of heredity: what it was, how it worked, how it was encoded in the newly named genes. To embark on this quest, you didn't need to be a physiologist or indeed a biologist of any kind. You needed to be a chemist or a physicist.[‡]

Along with this came a sea change in how physiologists looked at the phenomenon of homeostasis. Where Bernard had proposed a radical insight into the nature of life itself, his successors saw mainly mechanism to be teased out. Now drained of its vitalist core, Bernard's idea became a dried husk, roped in, tamed, and enervated.

It is somehow fitting, therefore, that it was physicists and engineers, not physiologists or biologists, who did the most to carry the

---

[*] Which he first used in a festschrift for Bernard's colleague, the Nobel Prize winner Charles Richet.

[†] The concept of the gene as an "atom of inheritance" was advanced by Hermann J Müller (1890–1967), one of the "rediscoverers" of the work of Gregor Mendel and one of principal figures in developing Mendelian genetics in the early twentieth century. This history is considered in more detail in Chapter 7. For a fuller history of the concept, see Falk's *Mendel's Impact: A Cultural History of Heredity* (University of Chicago Press: 2012), 9–30.

[‡] Some physicists are rethinking Rutherford's infamous dismissal, referenced above, of scientific "stamp-collecting," including an acknowledgment of how physics can, in fact, be informed by, say, sociology. See Hayes's "Undisciplined Science," *American Scientist* 92 (2004), 306–10.

idea of homeostasis forward into the twentieth century. As is often
the case, the needs of war provided the impetus. The technological
transformation of war through the early twentieth century is a well-
known story, as machines, not human beings, increasingly became
the means of war. Traditionally, war had been motivated by the
intelligence and bravery of warriors acting alone or en masse, and by
the tactical and strategic genius of generals;[12] but war in the twentieth century became a contest between brute machines, transforming the valiant warrior into meat for a cruel and impersonal grinder.
Wars still had to be won or lost, though, and military visionaries realized this meant that the ineffable human element of war had to be
brought back into control over the war machines' brutal lethality.[13]
So, entirely new kinds of mechanized war needed to be developed.
No longer was it sufficient that a machine could apply brute force to
an adversary. Rather, it had to apply force with the same intelligence,
tactics, and strategy of the warrior, and do it faster, more accurately,
and with less risk to living warriors and their comrades.

The great Norbert Wiener (1894–1964) of the Massachusetts
Institute of Technology laid the groundwork for realizing this vision,
in machines that ranged from the mundane (like a self-guided tor-
pedo) to the sublime (like a computer).[14] What came out of this vision
was the clockwork homeostasis: the homeostasis machine that did
the deed, but with its vital core safely reamed out.

Like Bernard, Wiener drew his inspiration from the seemingly
universal tendency of living systems to self-regulation, in forms
ranging from simple maintenance of an internal property like body
temperature, to goal-seeking behavior, to intelligence and learning.
Where Bernard saw a fundamental property of life, though, Wiener
saw a machine at work. If a machine could be made that could mimic
life's machinery of homeostasis, that machine would behave as if it
were homeostatic, that is, self-regulating. And that would be good
enough.

Wiener's early World War II work on self-aiming antiaircraft guns illustrates the problem admirably. Wiener described the problem in this way:

> The antiaircraft gun is a very interesting type of device. In the First World War, the antiaircraft gun had been developed as a firing instrument, but one still used range tables directly by hand for firing the gun. That meant, essentially, that one had to do all the computation while the plane was flying overhead, and, naturally, by the time you got in position to do something about it, the plane had already done something about it, and was not there.
>
> It became evident—and this was long before the work that I did—by the end of the First World War, and certainly by the period between the two [wars], that the essence of the problem was to do all the computation in advance and embody it in instruments which could pick up the observations of the plane and fuse them in the proper way to get the necessary result to aim the gun, and to aim it, not at the plane, but sufficiently ahead of the plane, so that the shell and plane would arrive at the same time as induction. That led to some very interesting mathematical theories.[15]

At one level, then, the problem of shooting an airplane out of the sky is quite simple: aim a gun at a target so the projectile and target meet at the time of explosion. It is so simple an idea, in fact, that we scarcely give a thought to what goes into successfully connecting projectile to target. Information (the visual image of the target) must be analyzed for information about its physical location: its distance, its elevation, its trajectory in space. A device, the gun, must be aimed at it, so that a projectile launched from the gun will arrive at the target's future location and detonate there. Shooting down an aircraft is, in short, a problem of information flow and management:

information about the environment (the location and trajectory of a target) comes in and is used to inform the operation of a machine (the gun) that will change that environment (obliterate the target).

As long as aircraft were slow, low-flying, and fragile, a skilled artilleryman could do all this on his own, and early antiaircraft warfare was little more than that—cannon fire directed against stationary or slow-moving targets, like an enemy's observation balloons. For a mobile target, like a fighter plane or bomber, the problem quickly becomes formidable. Altitude, speed, and direction of the target and time of the projectile's flight to the target all have to be taken into account. As warplanes became higher-flying, faster, and more maneuverable, the speed of reckoning needed to place the projectile at the target quickly outstripped the capacity of human brains and human-operated machines. The obvious solution to this problem was to develop machines that could do the reckoning faster and to make the machine, not the gunner, operate the weapon. The artilleryman's own reckoning skill could then defer to the mechanical "brain" that actually aimed and fired the gun. Some progress toward this goal occurred during the First World War, but the perilous peace that followed added urgency to the endeavor, so that a furious arms race unfolded between development of ever faster and more maneuverable aircraft and the ever faster and more sophisticated automation of aiming, fusing, and firing projectiles to blast those aircraft out of the sky.*[16]

Square into this problem came Wiener, child prodigy, mathemat-

---

* A similar arms race attended to the opposite problem, namely, how to deposit a bomb on a stationary target from a moving platform. The solution that emerged was the famous Norden bombsight, developed largely by the émigré engineer Carl Norden. The Norden bombsight incorporated an analog computer that solved the trajectory of a bomb for a given altitude, ground speed, and wind velocity. In its earlier versions, the bombardier actually flew the bomber. In its later versions, the bombsight itself flew the plane on its bombing run. For a fascinating, if tendentious, telling of the story of the Norden bombsight, see Malcolm Gladwell's 2011 TED talk "The Strange Tale of the Norden Bombsight," at www.ted.com.

ical genius, polymath, and prophet of cybernetics—the realm of the self-directed machine. (The word "cybernetics" translates literally from the Greek *kybernetikos,* rendered as "good at steering" or "good pilot.") The outlines of Wiener's biography are astonishing: graduation from high school at age eleven, bachelor's degree in mathematics from Tufts University at age fourteen, Harvard doctorate in zoology by age seventeen—maintaining through all this a serious study in philosophy. This remarkable intellectual career underscores Wiener's most salient trait as an engineer: he was never content simply to find a practical solution to a problem, but was compelled to burrow deep into it, always to seek the underlying principles of engineering problems, indeed the very *philosophy* of the problem. During World War II, the engineering problem that consumed Wiener was the aiming of antiaircraft weapons, which, despite all the technological advances in ranging and tracking that preceded him, still performed dismally.

Wiener saw that the fundamental problem was not just one of ranging, or of better mechanical tracking, or of improved fuse timers or triggers; it was a problem of *information* and *uncertainty.* Information about a target—its distance, elevation, and trajectory—came into a tracking system. These attributes all carried certain rates of change and error. That information then had to be fed into a machine that calculated some probability of a specified future state, namely, where the target likely would be at the end of a projectile's flight. Once this analysis was in hand, the gun would have to track its aim to the target's less-than-perfectly-predictable trajectory and make a decision whether to launch the projectile or to wait for a bit in case more information later would produce a more favorable outcome.

What was needed, in Wiener's opinion, was a general theory and formal mathematics of this ongoing process, which came to be embodied in the simplest cybernetic system: the closed-loop

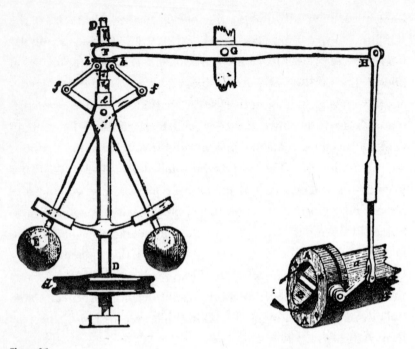

Figure 4.1
The Watt governor, or centrifugal governor, invented in 1788 by James Watt.

negative feedback controller. Wiener did not invent the idea of negative feedback control: the famous Watt governor for steam engines, also a negative feedback controller, had been invented a century before (Figure 4.1). Steam engines, however, were simple, stupid, and predictable, and this meant that the machines that controlled them, like the Watt governor, could be stupid, simple, and mechanical.* In contrast, a fighter plane or bomber steered by a pilot intent on a target was clearly intelligent, devious, and motivated in ways

---

* The Watt governor was a simple mechanical regulator of steam pressure at the turbine of a steam engine. The governor rotated in synchrony with the turbine and controlled a throttle valve for the steam. Two metal balls were attached to the governor's shaft on a hinged mount, so that faster rotation speeds extended the two metal balls outward. If the turbine spun too fast, the outward extension of the balls would throttle the steam, reducing its pressure and hence speed of rotation of the turbine, and vice-versa if the turbine was rotating too slowly.

that steam engines were not. This meant that machines that sought to shoot them down had to be intelligent in a way that a Watt governor did not have to be. Wiener saw clearly that a far more general theory of intelligent, goal-seeking agents needed to be developed. Wiener's approach to the problem—delving into its philosophical core—meant that once machines could be made to behave "intelligently," it would be possible to do much more than simply aim anti-aircraft guns. The way would be open to developing truly intelligent machines.

<hr/>

Wiener's closed-loop negative feedback controller had a particular resonance for physiologists. Indeed, it was Wiener's training in zoology that inspired him to think of them. So, it's worth delving into some of the details of how this type of cybernetic device works.

The closed-loop negative feedback controller consists of several subsidiary machines hooked together in a particular way (Figure 4.2). At the heart of the controller is a device called a *comparator.* As the name implies, this device compares information from two sources. The first, the *set point,* is an internal source that specifies a desired state of the environment, and the second, the *signal,* is information about the state of the environment as it is. In the example of a self-aiming gun, the set point might be to position the target squarely in the center of the gun's crosshairs, while the signal would be the actual location of the target with respect to the crosshairs.* The comparator takes these data and calculates an *error signal,* which encodes any mismatch between the set point and signal, in this instance, the

---

* During the interwar years, weapons developers tried a variety of schemes for feeding in information, including acoustic ranging, sightings by multiple spotters linked by telegraph, and so forth. Radar, however, quickly emerged as the best technology for this, and when war finally came, it was the British advantage in radar technology that was decisive in the Battle of Britain.

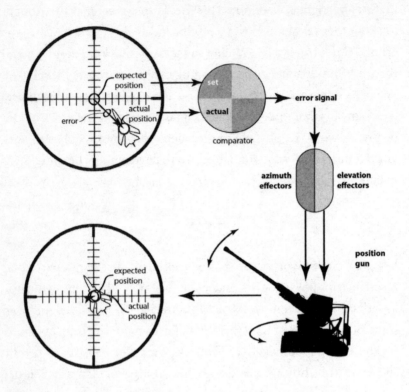

Figure 4.2
A closed-loop negative feedback controller.

deviation between where the aircraft is and where the gun is aimed. The error signal is then sent to a series of *effectors:* in a self-aiming gun, these will be the motors and gears that rapidly swivel the gun on its mounts to adjust azimuth and elevation angle. As the gun moves, the relative positions of the aircraft's image and the center of the crosshairs will change. This will generate a change in the signal, which will feed back onto the comparator, which can generate a new error signal, which adjusts the aim again. Thus, information flows in a continuous closed loop between comparator and effector, all engineered with the goal of negating any deviation between set point and signal, in this instance, between the position of the target in the sight and the center of the crosshairs. It is, in short, a *negative* feedback controller.

That's simple enough to grasp, and if everything is calibrated and working properly, the controller will continuously work to minimize the error signal, that is, to keep the moving target positioned squarely in the sight's crosshairs until the projectile is launched. It's the "if" in that last sentence that's the problem, though: what precisely do we mean when we say the system is calibrated "properly"? In fact, "proper" calibration is a delicate balance between sensitivity and response of the controller's various parts. If the system is too sensitive to perturbations in how it senses the actual state or if the response of the motors is too emphatic, the gun will swivel wildly, "hunting" all over the sky but never being able to zero in on the target. If the miscalibration is the other way, the launched shells will always end up at the place where the target had been rather than where it will be. This is the problem Wiener solved, setting down the theory for optimizing the performance of any closed-loop negative feedback controller. In so doing, he laid the foundational theory for all self-controlling mechanical systems.

Once Wiener systematized and mathematized the concept, the skies opened, so to speak. They were pried wide in no small part by Wiener himself, who was much more than just a mathematician and an engineer. He saw in the negative feedback controller, indeed in all such "well-steered" or "cybernetic" mechanical systems, the solution to the philosophical problem that had bedeviled biology for centuries, namely, how to explain life's obvious purposefulness without having to resort to the mysticism that it implied.[17] If living things were themselves cybernetic systems, then all that troubling philosophical baggage would fall away. Open now was the secret of life, of the mind, of the organization of society and economies, of politics, of health and illness. And in the years following World War II, which had been won just as much by those with the slide rules as by those bearing rifles and steering the tanks and ships and piloting the planes, plenty of people were ready to take up the challenge and to bring cybernetic technology to the service of the new world a-dawning.

Looking back on those heady first days of cybernetics, it's hard not to be swept up in the enthusiasm and technological optimism the field engendered. To get a glimpse of this brave new world, one could, for instance, take a stroll through the proceedings of the famous Macy cybernetic conferences and their intriguing titles—"The Algebra of Conscience" is my personal favorite. Remarkable personalities participated, including a stellar pantheon of the advanced thinkers of the day drawn from fields as diverse as computer science (Wiener himself), psychiatry and neurobiology (Warren McCullough), mathematics (Wiener's archrival John von Neumann), anthropology (Margaret Mead), behavioral and cognitive science (Gregory Bateson), genetics (Max Delbrück), and many more.[18]

But you can also get an idea of the climate by paging through any popular science magazine from the 1950s, such as *Scientific American* or *Popular Science*. In the advertisements, the articles, and the commentary, the vision was enticing: the door was opening to a new age and cybernetics was the key that would open it. Machines would now serve us, taking over the many drudgeries that consumed our everyday lives: self-operating machines would steer our vehicles on the roads, over the seas, and through the air; robots would stand tirelessly on assembly lines doing repetitive drudge work for months, not hours, at a time; machines would even do our thinking for us. And legions of men in short-sleeve white shirts and crew-cuts would take us there, guided by the genius of cybernetics.

It wasn't all sunny uplands, of course. Much of the technological boosterism that oozes from the pages of these magazines was fueled by military necessity. We were cheerfully told that even a missile could have a high IQ and home in on enemy "pigeons" like a stealthy falcon (Figure 4.3). Looking back on those days leads one to reflect on just how dangerous the world was then and how fortunate we are to have come through it intact. One could also find subtler but equally dark shadows draping other areas of the social landscape. When the cyberneticians turned their attentions to people and societies, for example, the enthusiasm and optimism that buoyed up smart missiles could easily lapse into bone-chilling arrogance.[19] Consider how the behaviorist B. F. Skinner put the prospects for engineering human societies along cybernetic principles: "All men control and are controlled. The question of government . . . is not how freedom should be preserved, but what kinds of controls are to be used and to what ends."[20] Batteries weren't included, presumably.

Figure 4.3
The cybernetic promise of the 1950s.

But, back to homeostasis. If cybernetics found its inspiration in biology's self-regulating systems, then physiologists in the 1950s began to look to cybernetics to return the favor: perhaps cybernetics could go beyond merely imitating homeostasis and solve the very problem of living homeostasis itself? And so was born the notion of homeostasis as the outcome of a negative feedback control machine. The clockwork homeostasis of Norbert Wiener and his many acolytes was adopted into the biological family.

⸻

Opinion will, of course, differ on how that's all been working out. On the one hand, remarkable insights and benefits have flowed from a cybernetic approach to difficult physiological problems. Certain types of movement disorders, like the tremors of Parkinson's disease, can be explained remarkably well as the "hunting" of a poorly tuned negative feedback controller, as I just described, and this has led to some very successful cybernetic-inspired treatments for such debilitating conditions.[21] Recent remarkable development of "smart" prosthetic appendages likewise owes its success to understanding the essential cybernetics of limb motion and incorporating it into mechanical limbs.[22] Likewise, the head-spinning recent advances of artificial intelligence, of computers that can learn to play Pong from scratch or that can win at the game show *Jeopardy,* were made possible largely by intelligence being treated as a cybernetic system of learning and feedback.

On the other hand, there have been failures, and the reasons for the shortcomings are instructive because they underscore the ultimate inadequacy of conceiving of life as if it were a machine. I want to illustrate this using a particular kind of clockwork homeostasis as my example: regulation of body temperature. I've chosen this in part because body temperature homeostasis happens to be the field

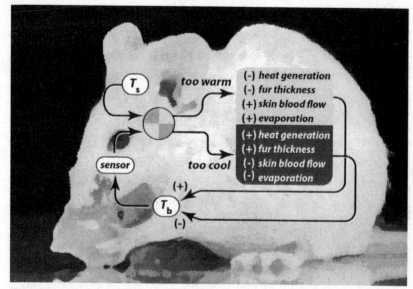

Figure 4.4
The living thermostat?

where I cut my professional teeth, but also because temperature homeostasis was among the first regulatory problems that physiologists came close to taming by cybernetics. It also illustrates how the machine metaphor for homeostasis can unravel.

Body temperature is clearly a regulated property. If an animal's body temperature exceeds some value, in ourselves about 38°C, heat loss mechanisms like sweating or panting or flushing of the skin are activated and heat generation is dialed down. If the body cools to a temperature lower than this, heat retention and heat generation mechanisms, like shivering or withdrawal of warm blood from the skin, are activated (Figure 4.4). The result is a body temperature that is steady as she goes, sustained at this target body temperature, the *set point* temperature, even as environmental temperatures and circumstances vary widely.

Armed with the metaphor of the clockwork homeostasis, physiologists in the 1950s began the search for the components of the body's living thermostat. For a time, the quest yielded great dividends. The first breakthrough precisely located what seemed to be the brain's "thermostat." By penetrating the brains of sheep and dogs with tiny probes that could heat or cool local patches of brain tissue, physiologists quickly located the putative thermostat in a small region at the base of the brain, just in front of the pituitary, known as the preoptic anterior hypothalamus, or POAH.[23] By heating this patch, the brain could be tricked into thinking the body temperature was too high and the body would anomalously begin to dump the supposedly "excess" heat. By cooling the patch, the trick could be reversed: the brain would think the body was cooler than it actually was and would direct the body to conserve heat and stoke up the body's furnaces.

Once the POAH thermostat was found, the other putative components of the brain's negative feedback controller started turning up. Certain neurons in the POAH, for example, would fire at a rate proportional to POAH temperature; these had to be temperature sensors, encoding temperature in their firing rates. Other POAH neurons would fire at a steady rate, no matter what the temperature; these had to be the sources for the set point. To add icing to the cake, the set point even seemed to be adjustable, like the thermostat of a house. The puzzling problem of fever, for example, was shown not to be a breakdown of thermoregulation, as physicians had long thought it to be, but a simple upward adjustment of the brain's "thermal set point."[24] Just as house temperature could be made to rise by turning the dial up on the thermostat, so too could body temperature be made to rise or fall with the adjustment of the brain's thermostat. It all made such satisfying sense. Thermal homeostasis was solved.

Except . . . the clockwork homeostasis model for body temperature regulation actually has not stood up well, not because it was unsuccessful, but because its success was always imperfect. Thanks to Norbert Wiener and his formulas, one could, in principle, predict the behavior of any cybernetic system from the system's measurable

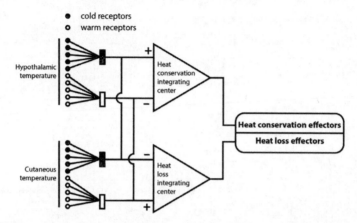

Figure 4.5

An early cybernetic model of temperature regulation, elaborated to include inputs of skin temperature with brain temperature in the preoptic anterior hypothalamus, feeding into "blackbox" integrating centers for heat loss and heat conservation.

parameters of operation: the sensitivity of the sensors, the responsiveness of the controllers, the time lags between sensing and response. If body temperature was controlled by a cybernetic controller, then Wiener's mathematical formulas should enable one to calculate the specific response of body temperature to known perturbations. That opened the door to some real science: one could predict the behavior and see experimentally how close the agreement was between predicted behavior and reality.

Invariably, these experiments showed encouraging agreement with the cybernetic model, but also some significant departures. What, then, should a "real"—that is, a reductionist—scientist do? Do we question whether the beautiful model is a reliable facsimile of the real system at work, or do we tweak the model to make it behave "properly," that is, more like the observed behavior? Invariably, tweaking the model was the preferred choice, which meant generating a more complex cybernetic model that, it was hoped, could more precisely model the actual behavior (Figure 4.5). And again, invariably, these new models produced some agreement, and new

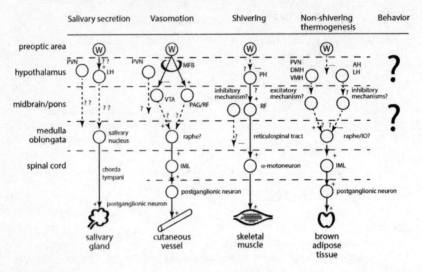

Figure 4.6
A more complex modern model of the clockwork homeostasis for body temperature regulation. The details are less important than the *gestalt*: the complexity, the large numbers of question marks, and the dotted lines with arrows that indicate an uncertain neural pathway.

discrepancies, which necessitated more complex models, which led to new discrepancies, and . . .

You get the idea. This endless tweaking and fixing of the purported cybernetic machinery of thermoregulation proliferated into baroque assemblages of multiple controllers, sensors, and effectors distributed throughout the nervous system and body, all arising ad hoc as explanatory needs multiplied (Figure 4.6).[25] If this sounds familiar, it should: it is the same motivation that led to the eighteenth-century proliferation of vital essences that so perturbed François Magendie. Now, one can convincingly argue that this is nothing to be troubled over. Are not organisms extraordinarily complex? Would not we then expect their cybernetic control systems to be commensurably complex? And is it not perfectly natural that the "proper" scientific trajectory should be to ever more complex cybernetic models until they ultimately match the complexity of the thing being modeled? For the scientist working today, this logic is sound and compelling.

But models can be a siren song, particularly if the model is beautiful, and this can lead scientists into an unreal realm where the model supersedes the thing it seeks to explain.

There certainly is historical precedent for this, for it was this very cycle of seduction, disappointment, and requital that drove and sustained for centuries the unreal Ptolemaic picture of the geocentric solar system. Slight discrepancies in the motions of planets and stars could be fixed by just adding one orbital epicycle *here,* but that's not quite right, so just another epicycle *there* will fix the problem, and so forth and so on. These accumulated fixes and tweaks made for a remarkably accurate model for the motions of the heavenly bodies, even as they made the Ptolemaic model physically absurd.[*][26]

Similarly, the search for the clockwork thermostat has led to ever more complex and sophisticated cybernetic models, until the clockwork thermostat finally converged on the ultimate complexity of the "many little lives" metaphor. Rather than nerve cells and sensors serving as dedicated components in neural "circuits" that are wired together into a well-tuned cybernetic computer, there are innumerable little conversations, spreading of gossip and vague rumors, and the "weather" inside the brain. Individual nerve cells seem to be homeostatic agents unto themselves, capable of sensing temperature, making comparisons, and even bringing about a degree of self-maintenance of temperature on their own.[27] Just as Henderson and Cannon had been led by their studies back to the "many little lives" metaphor, there too has the cybernetics of thermoregulation been led. There is no master thermostat; every cell is a "-stat" of some sort. The steady temperature of the body somehow emerges from the endless conversations and negotiations between these innumerable many little "-stats."

---

* It's a curious fact of history that it was not accuracy that buoyed the Copernican heliocentric system but Kepler's appeal to the five Platonic solids as the philosophically "right" model for the motions of the planets.

The metaphor of the clockwork homeostasis runs into deeper trouble when it ventures beyond the sheep and dogs that served as the early "model" organisms for cybernetic body temperature regulation. Creatures such as these were well-suited as experimental "test beds" for exploring the clockwork thermostat, because they have in common one style of temperature homeostasis, namely, high and steady body temperature that is controlled by high rates of internal heat production—the internal "fire of life" that, for a mammal like a sheep or a dog, burns off directly as heat as much as 90 percent of the energy in the food consumed.[28] This "style" of temperature regulation, called *endothermic homeothermy,* describes our own form of temperature homeostasis, so it's very important to us; but endothermic homeothermy actually is rare among the animals. It is found mainly among mammals and birds, but for the rest of the animal kingdom, a broad diversity of body temperature "styles" prevails, ranging from just drifting along with the temperature of the environment to actively seeking thermal comfort[29] to occasionally taking a kind of regulatory respite, as torpid animals do.

Among these diverse styles of thermoregulation is a body temperature that is determined not by internal heat generation, as it is in mammals and birds, but by the clever exploitation of external sources of heat. Creatures that regulate their body temperatures in this way were long called "cold-blooded," but this is a misrepresentation. Many lizards, for example, can maintain impressively high and steady body temperatures through the day.[30] Where a mouse in the cold morning might burn through its food reserves to warm its body, a lizard might spend the early morning hours sunning itself on a rock, capturing solar heat to offset its body's heat loss to the cool morning air. Once the lizard's body has warmed sufficiently, it may begin making forays into cool parts of its territory to find food, cooling as it forages. When it starts feeling too cold, the lizard might then find a sun fleck to soak up a bit of heat to warm its body, then go off into the cool to forage some more. By shuttling all day between hot spots and cool spots, warming up a little now, cooling a little then, again and again, the lizard can keep its body within a narrow band of temperatures.[31] In short, a supposedly "cold-blooded" lizard not only can be "warm-blooded," it can be remarkably adept at regulating its temperature *behaviorally.*

Behavioral regulation of body temperature poses some interesting challenges to the cybernetic metaphor for homeostasis. Like mammals, lizards also have a POAH in their brains, and as in mammals, the lizard POAH seems to play a role in the lizard's regulation of its body temperature.[32] The behavioral dimension of lizard thermoregulation throws a spanner into the cogs, however, and it's well-illustrated by the matter of fever. Lizards that have a bacterial infection also develop fevers, and their fevers are simultaneously like and unlike the fevers of endothermic homeotherms such as ourselves.[33] Whereas a febrile homeotherm drives up its body temperature by ramping up its heat production, a lizard develops a "behavioral fever" that is brought about by adjusting its shuttling schedule so that it spends proportionally more time in warmer environments and less time in cooler environments than it would if it were healthy. Thus, the febrile lizard's body temperature oscillates around a higher mean value than a healthy lizard's does. So far, so good: we can imagine that the lizard has a "behavioral thermostat" that has been turned up by fever, just as in a febrile rat.[34]

There's a subtle problem with this analogy, though, because the notion of a cybernetic "controller" becomes blurred when the thermoregulation is behavioral. A cybernetic thermostat is an attractive

---

* The formal name given to this style of thermoregulation is *ectothermic poikilothermy.* "Ectothermic" refers of course to the animal's use of external sources of heat; "poikilothermy" refers to body temperature being "variegated."

metaphor when it is the physiology of the body that is being driven by the cybernetic machine. There, the cybernetic machine holds the whip in its hand—that's why it's called a *controller,* after all. When I have a fever, I have no choice in what my body temperature is. It simply is what it is; my body temperature is controlled by the machine running my body, set at whatever the machine says, irrespective of time of day, how I'm feeling, what I've recently eaten, and so forth. This is not the case with behavioral fever, nor is it, as we shall see, the case for behavioral thermoregulation in general. A febrile lizard has a high body temperature because, in some sense, it *wants* to have a high body temperature, and it acts on this desire in a way that is impossible for me to do. In what sense, then, is the lizard's brain "thermostat" driving its body temperature in the same way *my* brain "thermostat" supposedly does?

Things get fuzzier still when we explore how lizards maintain their body temperatures in the real world. There, body temperature is not just physiology; it is a broader ecological problem because the body temperature affects nearly all aspects of how the lizard fits into its environment: its adaptation, in a word. A lizard's body temperature affects how fast it can move, how big its territory is, how intelligent it is, how successful it is at catching prey, how effectively it digests its food, and how successfully it reproduces.[35] What temperature the lizard maintains is therefore the result of a careful balancing of benefits against cost.[36] Some of these are so-called opportunity costs. When a lizard basks on a rock, it is forgoing opportunities to catch food, to find mates, and to fend off rivals. There is also frank risk. Because good sun-basking sites are often well-exposed sites, this makes the lizards that use them conspicuously visible and susceptible to being snatched up by a sharp-eyed predator. What temperature a lizard maintains will therefore be determined by a balance sheet of sorts. If, for example, sunny spots are abundant, as they might be in an open desert, the risk of predation that accrues to sun-basking in any one spot is low: every place in an open desert is equally visible, so there is small additional risk to basking in a particularly nice spot. Opportunity costs might be similarly low: a lizard lying on a rock in an open desert landscape can easily keep a watchful eye out for potential morsels of food strolling by while it basks and it can easily pounce when something juicy turns up.

The balance sheet can change quite a bit for lizards living in more sheltered environments. If you live near some woods, take a walk through them and look around as if you were a lizard looking for a place to bask in sunlight. In a typical forest, suitable basking spots are limited to the scarce flecks of sunlight that stream between gaps in the forest canopy. Now try to take the lizard's-eye view of how to maintain a particular body temperature. The scarcity of suitable basking spots boosts the frank risks to a basking lizard because this makes any prospective predator's job easier. If the predator wants lunch, it just needs to find a sun fleck, hang around there, and the probability will be high that a tasty lizard will take the chance and try to bask there: dinner is served. There are also increased opportunity costs. When sun flecks are scarce, the lizard must leave a large territory unpatrolled while it takes time out of its day to bask.

This leads to an interesting prediction. In an open sunny environment, there are few costs to pay and small risks to take if a lizard living there wants to have a high body temperature. A lizard living in a forest where sources of heat are scarce pays stiff penalties to attain the same temperature. The actual body temperature a lizard maintains should therefore correlate with those costs, so that lizards in open and sunny environments should sustain higher and steadier temperatures than do lizards in more sheltered environments. At least for lizards inhabiting Puerto Rico, the prediction is largely borne out.[37] Furthermore, there appears to be little genetic variation that accounts for the different body temperatures maintained by lizards inhabiting different environments.[38]

It appears, then, that lizards actively take stock of their environments and determine what temperature they will sustain based upon a perceived matrix of costs, benefits, and risks. If the costs are low, then get that body temperature right up! You'll be faster, more agile, more intelligent, and you'll go through your meals more quickly. But if the risks are too high, the lizard will let its body temperature slide. Now, I have to make the obligatory disclaimer that I am not saying that the lizard actually *thinks* these things. The point I'm trying to make is that the body temperature of a lizard is not so much the outcome of a machine regulating it, but is a kind of cognitive state.[39] This is where things begin to get even more metaphysically hairy for the metaphor of the clockwork homeostasis, because it opens the door to that anathema of modern biology, *intentionality*.

In *The Tinkerer's Accomplice,* I argued that cognition and intentionality are flipsides of the same underlying phenomenon of homeostasis. Cognition involves forming a coherent mental image of the "real" world, and the coherence of that mental image depends upon a homeostatic brain. Intentionality is the obverse of this: intentionality is the reshaping of the real world to conform to a cognitive mental image. This also depends upon a homeostatic brain. In short, all homeostasis involves a kind of wanting, an actual desire to attain a particular state, and the ability to create that state. The nexus of strivings and desires in, say, a lizard might be completely alien and inaccessible to the strivings and desires in, say, ourselves, but they are no less strivings and desires all the same. A clockwork vision of homeostasis cannot ever hope to capture this dimension of the problem, because in what way can a thermostat "want" to achieve a particular temperature in the same way, say, a lizard might "want" to do the same?

———— • ————

We now come to the nub of the problem that will occupy the rest of this book. I mentioned in the first chapter that Darwinism, as Darwin himself conceived it, was primarily a theory of evolution driven by adaptation. Without adaptation, natural selection cannot work. Darwinian evolution therefore relies upon a coherent theory of adaptation.

As we have seen, physiology already has in its pocket a robust theory of adaptation in the living organism. It is the Bernardian concept of homeostasis, the self-driven tendency of living things toward apt form and function. The question we address for the remainder of this book is whether the physiologist's conception of adaptation can be useful to modern Darwinism's conception of adaptation. The prevailing answer to this question among evolutionists has long been "no." Adaptation is certainly relevant to the Darwinian idea, so the argument goes, but it is not adaptation per se that is important, but *heritable* adaptation. Both physiologist and evolutionist may use the same word—adaptation—but they mean very different things by it. When we are talking about physiology, or day-to-day life, or adaptation in the physiological sense, we can tentatively speak of purpose and desire and still remain on the right side of intellectual respectability. We welcome as fellow scientists the practitioners of psychology, behavior, sociology, and economics—disciplines we sometimes indulgently call "soft" science, compared to the "hard" science (i.e., strictly mechanistic) many biologists yearn to practice.

It's a different story altogether when it comes to the problem of *evolutionary* adaptation. Speak of purpose and desire for evolutionary adaptation and you'll be quickly lumped in with God-botherers and other intellectually malodorous tribes. There's a reason for this: the physics envy that permeated biology in the early twentieth century has, among other things, driven a thick wedge between physiological adaptation and evolutionary adaptation. This is the fractiousness that comes with epistemic closure: it is one reason why we do not presently have a coherent theory of adaptation, nor a coherent theory of life, nor a coherent theory of evolution.

The path to a coherent theory of life, and hence a coherent theory of evolution, therefore depends upon there being a coherent theory of adaptation. As we have seen, physiology has such a theory but it contains at its core a difficult-to-digest nut. Adaptation in the physiological sense is really a phenomenon of cognition, striving, and desire. Living systems have to be aware of their surroundings, to be aware of what they are, and to strive, like the imaginary Mukurob of Chapter 2, to a particular state. This is the radical implication of Bernard's dangerous idea.

Modern Darwinism rejects this solution to its adaptation problem precisely because it contains those troubling concepts of purpose and desire. Modern evolutionism rejects this solution, not because it has disproved it, but because it is philosophically inconvenient. The virtue of natural selection is that it reduces evolution to the operation of a machine, and it is impossible to attribute purpose and desire to mere machines; therefore, there can be no purpose and desire in evolution, as the Four Horsemen of the Evocalypse have instructed us.

Yet, Darwinism, and evolutionism in general, did not always have trouble with the idea of evolution as a purposeful, striving phenomenon. As you might suspect, there is a narrative afoot intended to make us think otherwise.

5

A MAD DREAM

Enduring brutal winters is a point of perverse pride in upstate New York, where I now live. Since I grew up in coastal California, winter's charms are mostly alien to me, but I am certainly not immune to the charms of winter's end. After months of lowering gloom, the sunlight suddenly just seems a little more . . . optimistic. There's the first day you can leave the house unburdened by the heavy winter coat and the several layers of insulation you have to carry. Taking a deep breath outside suddenly no longer feels like shooting back a slug of white lightning but more like savoring a snifter of fine cognac, with the air redolent of the hebetic aroma of soil suddenly unlocked from winter's icy embrace. The experience is all the more intense for having endured the winter, which I suspect is the real source of the pride, a pat-on-the-back for all the shoveling and salting and deicing that are now over, at least until the next bout with winter, queued up to start again in about six months' time.

The conventional history of the rise of Darwinism—there's always a conventional history—tells a similar tale, of a bright spring, full of promise, emerging from a long, dark winter. Here is how one historian of Darwinism, Michael Ghiselin, pithily describes it:

[Darwin's theory of evolution] focused attention on the principle of natural selection, which a lot of people did not like. Some of the people were French, and their attitudes were colored by chauvinism. They wanted an explanatory principle devised by a Frenchman. Others rejected natural selection because they did not like the relentlessly competitive world that selection implied and that Darwin envisioned. They wanted something that would be more in keeping with their personal, beneficent political values. For these and other reasons, people espoused various alternatives to natural selection, and a number of factions were formed.

By the beginning of the 1900s, two of the factions had gained major importance and were dominating the debate about evolution. One faction comprised people who favored natural selection as the chief mechanism of evolution, and who rejected the inheritance of acquired characteristics; these people came to be called "Neo-Darwinians." The other group advocated a wide variety of evolutionary mechanisms (known or unknown), and they also accepted the premise that acquired traits could pass from one generation to the next; these people were called "Neo-Lamarckians."

The controversy between the two groups endured for several decades, but by 1940 biologists had learned so much about genetics and related subjects that the ideas of the "Neo-Lamarckians" were generally abandoned. They play no role in our modern theory of evolution, which emerged during the 1940s and the early 1950s.[1]

And there you have it: before Darwin, and for some time after, evolutionism was haunted by the ghost of someone named Lamarck, motivated by morality, idealism, and French chauvinism. What's worse, the Lamarckians were spirit-mongers, idealists who wanted

Figure 5.1
Famous figures of the "French Evolution." *(left)* Lamarck (1744–1829); *(right)* Cuvier (1769–1832).

to reinfect evolution with their discredited vitalist yearnings. They were doomed to lose to the Darwinists, who were the "real" scientists and who let facts, not wishes, speak with the loudest voice. Reason triumphed, and the crisis of Darwinism was solved.

But what of this fellow Lamarck? By the early nineteenth century, so the story goes, the idea of evolution was "in the air," put there by the evolutionary "speculations" of the French naturalist Jean-Baptiste Pierre Antoine de Monet, Chevalier de la Marck, or as he is better known, simply Lamarck (Figure 5.1).* Despite seeing the reality of evolutionary change, Lamarck went wrong because he believed that evolution worked through the "inheritance of acquired characteristics."[2] To trot out a well-worn example, giraffes had long necks because their okapi-like ancestors had long been stretching their necks to reach high up into trees for tender foliage. The stretching "took," and this is how the short-necked ancestors of giraffes

---

* It was Lamarck who coined the term "biology," to distinguish the study of life from the broader rubrics of "natural philosophy" and "natural history."

came to be the long-necked giraffes of today.* When we tell this story to students today, it is usually accompanied by a good laugh at Lamarck's expense: how silly an idea!

Another Frenchman is lurking in the wings of Darwinism's great saga. Following Lamarck was the great anatomist Georges Léopold Chrétien Frédéric Dagobert, Baron Cuvier, his name, again, and mercifully, condensed to Cuvier (see Figure 5.1). Cuvier is supposed to have reconciled his essentially creationist bent with the record told by fossils, the "formed stones" that Cuvier had, by his own hand, organized and crafted into what we now call the "fossil record." The story told by that fossil record challenged the origin story told in Genesis. There seems to have been not one creation, but many, each one different from the last. The history of life on Earth was marked by long periods during which animals and plants persisted unchanged, punctuated by episodes of catastrophic extinction, followed by proliferation of new species, when life settled into the next long period of stasis. According to Cuvier, we are living in the last of these serial creation events. Not only did this contradict the single creation of Genesis, it contradicted another dogma of the age, the conception of the unchanging species laid down by the great Carl Linnaeus (1707–1778). To Linnaeus, species were reflections of unchanging archetypes. The diverse starfishes, for example, were a reflection of an unchanging "starfishness" and so were unchanging themselves. The fossil record seemed to be saying otherwise: species changed, they had history, and they had beginnings and ends—quite radical ones at that.

Both Lamarck and Cuvier are held up today as avatars of a benighted past, who *almost* got to the gold ring of the "right" answer to evolution, that is to say, Darwin's answer, but were held back

---

* Lamarck also wrote of necks stretching in the other direction. The long necks of swans, he thought, were attributable to the birds straining to reach as deeply as they could in water to rich bottom sediments for their food. He also applied the same metaphor to the stretching of snakes' bodies.

by self-imposed constraints of vitalism and religious obscurantism. Lamarck is often credited with making the *idea* of evolution
respectable—putting evolutionism "in the air" that, like that first
breath of spring air, Darwin would eventually inhale and bring
to glorious fruition.* Cuvier, for his part, is credited with at least
two of the important place settings that would come to adorn the
Darwinian feast. The first was his challenge to the concept of the
unchanging Linnaean species, hence the title of Darwin's revolutionary book, *On the Origin of Species.* The second was the theory of
homology, which is the tendency of parts of animals to change and
take on different forms and functions.† Homology showed how lineages could transform from one form seamlessly into another: how
a fish's pectoral fins could transform into a frog's forelimb, into a
horse's front hoof, into the wing of a bird, into the hand of someone
playing Chopin. Homology gave Darwin the model he needed to
show how new species could arise, not catastrophically, but gradually and imperceptibly through minute changes acting over many
generations.

Despite this, both Lamarck and Cuvier are held to have fallen short
of the prize that Darwin claimed. Lamarck's intellectual failure was
twofold. First, he was fixated on something called the *scala naturae,*
literally, the ladder of nature, which was the supposed progressive
ratchet that led all life inexorably to greater and greater complexity and perfection, culminating, naturally, in humanity. Lamarck's
second failure was his obsession with an erroneous model of inheritance: the already-mentioned inheritance of acquired characteristics,
of which more momentarily. For Cuvier's part, he is supposed to have

---

* Although evolutionism was mostly a French obsession, it did migrate across the English Channel,
  most notably to Charles Darwin's grandfather, Erasmus Darwin.

† On the English side of the Channel, the theory of homology was developed more fully by the paleontologist Richard Owen, who came to be critical of Darwinism and natural selection as a mechanism
  of evolution.

missed out on the gold ring by being a creationist and catastrophist.[*]

    Like the folk stories that have accreted around Claude Bernard, this conventional history is flawed, being less history than a narrative that keeps out some uncomfortable ideas.[3] One uncomfortable idea is that both Lamarck and Cuvier are better understood in the context of *scientific* vitalism, not the vitalism of spooks and vital essences. This makes Lamarck and Cuvier more of a challenging proposition to the Darwinian narrative and not so easily dismissed with amusing stories of the long necks of giraffes. Even more uncomfortable is the light this shines on Darwin's own thinking, which puts Lamarck, Cuvier, and Darwin into a coherent and seamless alternate narrative of evolutionary thought—one driven not by blind mechanism, but by purpose and desire, and marked not by the differences between the three men, but by their similarities.

———————•—•———————

Both Lamarck and Cuvier rose from lowly beginnings to become giants of the French Enlightenment. Although they came to be bitter enemies, together they launched what we might call the French Evolution.[†] The relationship between the French Evolution and English evolutionary thought—perhaps we can call the latter the Glorious Evolution—was deep, complex, and like the revisionist history of Claude Bernard, built around a number of myths that we will have to sort through. One such myth is the caricature of Lamarck and Lamarckism as the bête noir of rational—that is,

—————————————

[*] Lamarck and Cuvier were bitter rivals, even after Lamarck's death: Cuvier outlived Lamarck, which enabled him to use his eulogy at Lamarck's funeral to berate his dead compatriot one last time. Their rivalry is often cast as a difference of opinion over evolution, but it probably ran much deeper than that, perhaps motivated by class and religious antipathy: Lamarck came from an impoverished line of aristocrats, and Cuvier scrambled his way up from the bourgeois; Lamarck was a soldier and war hero, and Cuvier was always a scholar; Lamarck was Catholic, and Cuvier was Protestant, etc.

[†] A hat tip to the psychologist James Brady, who coined the phrase.

Darwinian—evolutionism. Cuvier's creationist tendencies are a second. That Darwin rejected Lamarckism—got evolution "right"—is a third.[4]

Lamarck was the scion of a family of aristocrats from northern France who had fallen on hard times. The young Lamarck first sought his fortune, as so many in Lamarck's circumstances did, in military service to his king. At the young age of seventeen, his bravery during a skirmish in the Pomeranian War* won him a pension from his grateful sovereign. This bought Lamarck sufficient leisure to pursue a burgeoning interest in natural history, focusing in particular on plants. Lamarck was also a social climber, and he eventually worked his way into the directorship of the Jardins du Roi in Paris (renamed the Jardins des Plantes in the revolution's aftermath).

Cuvier, for his part, was born into a French Protestant family from Montbéliard, near the German border of the Duchy of Württemberg. After concluding a brilliant student career, Cuvier found employment as a tutor in natural history to a Protestant nobleman in Normandy. There, Cuvier stumbled into what we would now call a networking opportunity. At that time, France was in the midst of its painful postrevolutionary upheaval, which made life in Paris very dangerous. While on a stroll one day, Cuvier had a chance encounter with the physician and agronomist Henri Tessier, who was living incognito in Normandy. Because Tessier had been a favorite of the deposed king, he naturally had feared being caught up in the Reign of Terror if he stayed in Paris. When Cuvier saw through his disguise, Tessier initially thought he was doomed, but Cuvier assured Tessier that he had no intention of exposing him. His secret (and

---

* The Pomeranian War (1757–1762), between Sweden and Prussia, was part of the Seven Years' War. The seventeen-year-old Lamarck distinguished himself in an artillery barrage that reduced his company to about a dozen men. Once all the officers had been killed, the remaining men of his company prevailed upon him to take command, which he did, showing great courage that enabled his company to hold their position until relief and victory could come.

life) secure, Tessier and Cuvier became fast friends. Once Tessier was confident that he could return to Paris and keep his head, he took Cuvier with him, introducing him into the burgeoning intellectual life of postrevolutionary Paris. Once there, Cuvier was able to lever-age his connections and was asked to join the Académie des Sciences of the Institut de France. From his new Parisian base, Cuvier built his stellar career in paleontology.

Lamarck was older and a more traditional thinker than Cuvier, and so Lamarck's thinking was infused still with the search for essences and forces that could explain the organism's, and life's, unique properties of adaptability, coherence, and organization. Like his contemporaries, Lamarck's thinking was influenced by the growing conception of the organism as "many little lives," but he lived at a time when scientific vitalism had yet to outgrow its obses-sion with vital essences: the *vis mediatrix*, the vital stuff that coor-dinated it all, was still a viable idea. Lamarck therefore sought vital principles that could unify our understanding of the universe, from mineralogy to the weather to life itself, although he saw these more in terms of immaterial vital forces than vital matter. It's tempting to look at Lamarck's rather quaint essentialist ideas and summarily dismiss the rest of his thinking. Indeed, this was the very thing that Lamarck's opponents (beginning with Cuvier) have done through the years, and to great effect. This is a pity, because it paints a carica-ture of Lamarck that obscures his truly revolutionary idea.

Lamarck proposed that living systems are uniquely imbued with at least two vital forces. One is the *pouvoir de la vie* (life power), which is more accurately, if more cumbersomely, rendered in French as *la force qui tend sans cesse à composer l'organisation* ("the force that tends perpetually to make order," or perpetual order-producing force). In English, we describe this more succinctly as the "complexifying force," because it supposedly impelled living systems toward ever greater complexity. For individual organisms, the complexifying force is what drives, say, the transformation of a formless mass of yolk, semen, and egg into a complex living organism.

Lamarck's second vital force was the *influence des circonstances,* cast more simply as the "adaptive force." Within the individual, adaptive force modifies the body to help it meet changing and unpredictable environmental circumstances. If fur becomes thicker in the winter, if skin becomes darker under the sun, if muscles become stronger under strain, this is the adaptive force at work. Both the complexifying and adaptive forces were the mainstays of the vitalist thinking of Lamarck's day, and his thinking in this regard was not original with him. As we have just seen, Lamarck's complexifying force was essentially drawn from the same philosophical well as the epigenetic force that was said to drive embryonic development from the formless egg into the complex organism.*

Lamarck's real intellectual innovation was to propose that these forces could apply to *lineages* of organisms as much as they applied to individual organisms (Figure 5.2). Not only were developing embryos, for instance, marked by an ever increasing complexity through time, the lineages of successive generations of embryos were as well. Thus, the complexifying force could act both across and within generations. Similarly, if the adaptive force could be invoked to fit an organism more aptly to a new set of circumstances, so too could the adaptive force operate over many generations to bring about an ever more apt fit of a lineage to, say, a long-standing change of environment. If burrowing shrews lost their eyes to become moles, if shrews grew wings to become bats, if panthers grew lithe and supple to become cheetahs, this was the adaptive force working across generations. Put them both together, and you have a theory of

---

* In Lamarck's time, there were two principal ideas about embryonic development. Preformationism held that development was the unfolding of an already completely formed adult within the egg or sperm, called a homunculus. This form of the word "epigenesist" is quite different from the epigenetics of modern times, which is focused on physiological modifications of genes.

Figure 5.2
A Lamarckian phylogeny. Complexity increases with time, owing to the continuous operation of the complexifying force. Each lineage in turn increases in complexity over time, owing to the same complexifying force.

evolution: of lineages becoming ever more complex and ever more well-adapted through the lineage's history.

This puts Lamarck's thought into a different perspective than our accustomed view of it: it is not a mish-mash of rather silly ideas about giraffes stretching their necks or sons of blacksmiths having stronger arms than their fathers. To the contrary, take away the essentialist gloss, and we see some rather modern thinking on Lamarck's part. The complexifying force, for example, bears some similarity to the modern notion of spontaneous order emerging from open thermodynamic systems.[5] Lamarck's adaptive force, for its part, looks a lot like homeostasis, properly understood. The main point about Lamarck, though, is that his thinking was not, as the caricature would imply, centered on a wrong-headed theory of inheritance. Rather, he was proposing a radical connection between adaptation in individuals and adaptation in lineages. It was, for its time, a unified field theory of adaptation. This brings us to the reason why Lamarck has been the object of such caricature: his idea of a unified theory of adaptation is inconvenient to the modern narrative that physiological adaptation and evolutionary adaptation are completely separate things, as outlined in the last chapter. Lamarck begs to differ: they are one and the same.

If Lamarck was the grand theorist of the French Evolution (or system-builder, as he was then contemptuously called), Cuvier was its law-giver. He is sometimes portrayed as an anti-evolutionist, but this is a canard. Of course Cuvier was an evolutionist: his paleontological work built the best case then extant that species changed through time. What Cuvier rejected was not evolution itself, but the idea that evolution was gradual or operated through mere mechanical processes. The evidence from the fossil record clearly showed that life changed abruptly, not gradually, as Lamarck's model seemed to predict.

Again, we see elements of some surprisingly modern thinking here. The doctrine of punctuated equilibrium, which had a burst of popularity in the 1980s, emphasized patterns in the fossil record that were very similar to Cuvier's serial catastrophes: a species would persist unchanged for long periods, punctuated with bursts of rapid species proliferation. Punctuated equilibrium wasn't quite Cuvier's catastrophism, of course, but it affirmed Cuvier's reading of the patterns of the fossil record.[6] Catastrophism, meanwhile, had new life breathed into it with the growing realization in the 1970s that an asteroid collision with the Earth sixty-five million years ago may have hurried dinosaurs (and a lot of other creatures) off the stage.[7] Now we look for evolutionary catastrophes everywhere, with intriguing but mixed results. There is, for example, the thought-provoking idea of Nemesis, the distant Death Star whose gravitational pull periodically flings asteroids toward Earth, raining down death and extinction like Zeus's thunderbolts, revealed in the periodic waves of mass extinction that mark the fossil record.[8] Or perhaps these are just statistical noise. Who knows? Cuvier's catastrophes may have to be taken seriously after all, but that is a task for another day.

It is Cuvier's work in comparative anatomy that is most of interest to us, because it was thoroughly steeped in the emerging doctrines of scientific vitalism and in the same "many little lives" metaphor that so strongly influenced Claude Bernard. Comparative anatomy compares the forms of different species with one another: how a horse's hoof is like a mouse's hand, for example. Comparative anatomy is something we force modern students of biology to endure because it teaches them valuable lessons about homology. Homology is important to the Darwinian idea because it bolsters the empirical case for the gradualist model of evolution that Darwin championed. To Cuvier, homology was something different entirely—a model for how lineages were *resistant* to change through time. For Cuvier, comparative anatomy provided the evidence for his own theory of adaptation, which he called the "conditions for existence."[9]

This is a term of some confusion, which must be cleared up. Darwin appropriated this phrase as a metaphor for the organism's adaptive milieu, that is, adaptation to *ambient* conditions. But Cuvier coined the phrase to mean those conditions *within* an organism that would serve apt function. His logic went something like this: to survive and reproduce, an organism had to be objectively capable of surviving and reproducing; to do that, all the parts of the body had to work well together. These were Cuvier's conditions for existence. If the parts did not work well together, the conditions for existence would not be met, and debilitation, death, and failure of the lineage would be the outcome.

An enduring legacy of this logic was Cuvier's famous theory of the correlation of parts,* to which the modern science of allometry traces its roots.† The correlation of parts was the operational system for the conditions for existence: the meat on the bones, so to speak. If all parts of an animal had to function well together to meet the conditions for existence, a change of one part by itself would violate those conditions. There would therefore have to be specific *correlated* changes of multiple parts to ensure that the organism's conditions for existence could be met as the body changed. Furthermore, not just any correlation would do. If the upper limb bones—the femur and humerus—of, say, leopards elongated over time to enable their descendants to run faster and become cheetahs, a correlated *reduction* of the length of the bones of the lower limb would not help the leopard's cheetahlike descendants to run faster: the conditions for existence for fast running would not be met, because the total lengths of the limbs would be unchanged (Figure 5.3). Only a correlated elongation of *both* upper and lower limb bones would meet the conditions for fast-running cheetahs to exist.

This notion fits neatly into Cuvier's interpretation of the fossil record as intermittent bouts of catastrophic extinction, interspersed with long periods of stasis in which the form of lineages would not change. Cuvier took this as evidence of how difficult it was to change from one set of conditions for existence—those that made leopards possible, for example—to another set of conditions for existence— those that made cheetahs possible. One would expect, therefore, to

---

* The correlation of parts idea helped establish the phenomenon of homology, developed by Cuvier's British contemporary Richard Owen, which was such an important piece of the evidence Darwin marshaled for his own evolutionary theories.

† Allometry literally means "other measure," which in modern terms refers to the scaling relationships between different parts of the body. Limb length varies with body mass, for example, by a particular mathematical relationship known as a power equation, of the form $y = a \times x^b$. This mathematical form is nearly a universal property of living systems, now applied to everything from simple correlation of form to the extent of animals' home ranges.

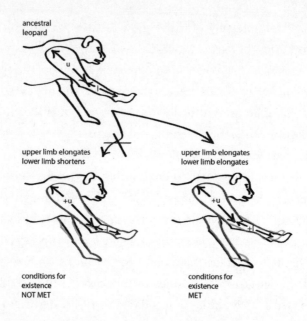

Figure 5.3
Cuvier's correlation of parts model for adaptation. For a leopard to be transformed into a fast-running cheetah, there must be a coordinated elongation of both the upper limb (u) and the lower limb (l).

see these correlations in the fossil record as generations of leopards became cheetahs. It was only when things were mixed up precipitously at an extinction catastrophe that a new set of conditions for existence among the survivors could begin to be negotiated and settled upon. Once settled upon, it was difficult to nudge lineages out of their new accommodations, sustaining the long periods of stasis Cuvier had claimed in the fossil record.

Cuvier worked out these particular correlations in exhaustive and exquisite detail for a variety of animals, both living and extinct, to the point that he could boast that he could reconstruct an entire animal from a single tooth. The shape of a tooth was related by a particular correlation to the jaw. Know the tooth, and you will know the jaw. The jaw, for its part, was correlated in a particular way to the skull, which was correlated to the spine . . . and so on. Know the tooth, and if you know the correlations, you will ultimately know the tiniest hairs on the tip of the animal's tail. Behind the braggadocio, there was an important point to Cuvier's paleontological prestidigitation: the correlation of parts is a clear example of the organism's "many little lives" working out their mutual accommodation within the context of the well-functioning organism. Indeed, you can see in Cuvier's thought the seeds for Bernard's own speculations about the constancy of the *milieu intérieur:* they are drawn from the same philosophical well.

There's another similarity that has bedeviled our understanding of Cuvier's evolutionary thought. Just as Bernard saw homeostasis as a fundamental property of living things, Cuvier saw the same in his conditions for existence. The specific correlations that had to exist for the animal to *be* well-adapted betokened a fundamental self-knowledge that permeated the organism. All the parts had to *know,* in a deep sense, how to fit in with one another, and to be *capable,* again in some deep sense, of working out an accommodation with the other parts that carried their own sense of knowledge and striving toward a destiny. At root, the "many little lives" metaphor was a statement of life as fundamentally a cognitive and intentional phenomenon. Purpose and desire, in a nutshell.

———•———

What of Darwin? Did he not leave all that quaint vitalist theorizing behind? Well, not really . . .

It's worth remembering that Darwin spent his formative years in an intellectual milieu that was similar to Lamarck's and Cuvier's and that was animated by the central question of nineteenth-century scientific vitalism: how did the well-constructed, coherent, and adaptable organism come to be? The oft-cited metaphor of evolutionism being something "in the air" that wafted over the English Channel from France, a spirit that Darwin just inhaled and brought to fruition, paints Darwin as more of an intellectual blank slate than he actually was. His novel thinking has to be understood in the broader context of nineteenth-century debates over the nature of adaptation.

We have seen the tumult that was roiling the medical faculties of Europe over this question. On the English side of the Channel, the notion of a marvelously designed living world had long planted its foot in the English school of natural theology, championed by a long line of thinkers from John Ray to William Derham to Thomas Malthus and culminating in the vivid apologetics of the great William Paley (1743–1805).[10] Darwin knew this tradition well but came into it sideways, as he did with so much of his thinking. He had come to know and admire Paley's thought and work when he was ensconced as a young man at Christ's College, Cambridge.* But he arrived at this point only after a circuitous path.

Charles Darwin was part of an extended family of free thinkers and liberals, part of the new professional elite that grew up during the Industrial Revolution.[11] He was born in Shrewsbury, in Shropshire, along the northern border with Wales. This area had seen considerable industrial development through the eighteenth century, although William Blake's "dark Satanic Mills" were less in evidence here than they were elsewhere in England.† Instead of

---

* Christ's had been Paley's home college at Cambridge.

† Blake described the Industrial Revolution's "dark Satanic Mills" in *Jerusalem*, a poem he wrote for the preface of his *Milton, A Poem* (1808). It takes its inspiration from a legend of Jesus visiting England in the company of Joseph of Arimathea. The "satanic mills" reference is found in the first two verses of *Jerusalem:*

> "And did those feet in ancient time,
> Walk upon Englands mountains green:
> And was the holy Lamb of God,
> On Englands pleasant pastures seen!
> "And did the Countenance Divine,
> Shine forth upon our clouded hills?
> And was Jerusalem builded here,
> Among these dark Satanic Mills?"

Blake was decrying the dehumanization and despoliation of England and its people by the disruptions of the Industrial Revolution.

belching coal, the clean waters of the River Severn provided most of the power that turned the wheels of the local textile mills. Shropshire was also blessed with extensive deposits of fine clay, which provided the raw materials for the fortune that elevated the Darwins into the upper middle class. This was the fine china produced by the Wedgwood family, whose family tree was intertwined closely with the Darwin's. But where the Wedgwoods became industrialists, the Darwins became professionals—physicians who could serve the growing numbers of people streaming into the towns from the countryside. Indeed, Charles was, for a time, destined to follow his father and grandfather into the family profession and was duly sent off to study medicine at the University of Edinburgh, where his father and grandfather had studied. There, Charles soon found he could not stomach the dissections and suffering that were part of his medical curriculum, and he left Edinburgh after one year. Exasperated by his son's failure at medicine, Charles's father directed him to the other respectable station for diffident sons of the middle class: study at Cambridge to enter the Anglican clergy. It was there that Charles came to know Paley's *Natural Theology* (1802), which was de rigueur for aspiring clerics, for it offered irrefutable proof of the existence of a wise, beneficent, and all-knowing deity. Paley was also important reading for students of natural history, toward which Darwin, with his obsession for beetles, shooting, and rat-catching, was drawn as if by gravity.

The fertile ground of Darwin's mind had been long prepared more by his family's profession of medicine rather than by natural history. Charles's grandfather, Erasmus, had studied medicine at Leiden, which was involved in the tumult over the transformation of vitalism that was roiling the eighteenth-century European schools of medicine. Consequently, there were many close connections between the medical schools at Leiden and Edinburgh. Charles's father, Robert, had also studied medicine at both universities, and he certainly would have been familiar with the doctrines of scientific vitalism that were emerging there.

Much of the history of evolutionary thought, and of medical thought for that matter, can be understood as an ongoing argument between two competing visions of nature: Romantic idealism and Enlightenment rationalism. Vitalism, and the insistence on the ineffable nature of the organism, was a reflection of the Romantic, as was the whole notion of natural theology as it applied to natural history. The English tradition of medicine, on the other hand, tended to lean toward the rational: the great physician William Harvey was English, after all, and Cambridge University was the font of Archibald Pitcairne's "Newtonian medicine," which sought to organize medical practice as a set of mathematical axioms.[12]

The tension between Romantic idealism and Enlightenment rationalism twined itself through the Darwin family tree. Robert Darwin was by all accounts a stolid, responsible, hardworking practitioner of practical medicine, by which he grew wealthy through assiduous attention to the everyday details of sickness and health. Erasmus Darwin, on the other hand, was a noted free thinker and admirer of the more Romantic tendencies of the voluble French. Both tendencies intertwined in Charles's idyllic youth ranging through the Shropshire countryside. From his grandfather, Charles became familiar with the radical concepts of species transformation streaming in from France: Erasmus had speculated about evolution and natural selection in several of his works, including his epic *Zoonomia* (1796), as well as in poems such as "The Loves of the Plants" (1789) and "The Temple of Nature" (1791). Erasmus Darwin was also no friend of mechanism, as can be seen in his preface to *Zoonomia:*

> It happened, perhaps unfortunately for the inquirers into
> the knowledge of diseases, that other sciences had received

improvement previous to their own; whence, instead of comparing the properties belonging to animated nature with each other, they, *idly ingenious, busied themselves in attempting to explain the laws of life by those of mechanism and chemistry*; ... forgetting that animation was [life's] essential characteristic.[13] (emphasis added)

Physics envy was a problem, it seems, even in the early nineteenth century.

In the heroic Darwinian story, Charles left all that muddy vitalist nonsense behind as the clear sun of natural selection began to dawn in his mind. It is certainly true that Darwin had critical things to say about both Lamarck and Cuvier—he disputed Lamarck's emphasis on progressive evolution and thought Cuvier's catastrophism was more simply explained as gaps in an incomplete fossil record. Nevertheless, the two core ideas of Darwin's own theory bear the unmistakable stamp of nineteenth-century vitalist thought. Success in the "struggle for existence" boils down to apt function (physiological adaptation), which draws inspiration, albeit flawed, from Lamarck's "adaptive force" applied to lineages. Homology, cited by Darwin as strong evidence that lineages could evolve through gradual modification of existing parts, drank deeply from Cuvier's notion of the correlation of parts and the mutual accommodation of the organism's "many little lives."

Indeed, in the cold light of day, Darwin's solution to the problem of evolution—natural selection—was not even especially original. Erasmus had crab-walked up to the idea of natural selection several decades before his grandson claimed it as his own. There were also at least two rival claimants to the idea of natural selection who preceded Darwin by many years, perhaps because Erasmus had put the idea "in the air," as Lamarck and Cuvier had supposedly done for evolution.* Finally, there is that remarkable coincidence of "Darwin's" idea independently popping up in Alfred Russel Wallace's fever-addled head.†[14] This is not to gainsay Darwin's own achievement, of course. What made Darwin stand out was his supreme talents as a naturalist, his attention to meticulous detail in observation, and his habit of delving exhaustively into evidence. These habits transferred to paper make for tortured reading, but they allowed him to see the problem of evolution from a perspective that neither Lamarck nor Cuvier nor any of the other armchair claimants to Darwin's idea could see or even remotely begin to defend.‡ Where Lamarck saw striving for perfection and unrelenting progress and sought a "system" to explain it, Darwin looked to nature as it was and saw evolution's sometimes whimsical tricks—and saw that these deserved explanation too.[15]

Darwin also recognized the fatal problem lurking in any theory of evolution that did not somehow synthesize apt function with heritable memory. Lamarck had seen the connection but had failed to close the deal—he had said only that they might be combined somehow but had taken it no further. And this is why Darwin undertook

---

* The two claimants were Robert Chambers, a Scottish journalist whose anonymous 1844 book *Vestiges of the Natural History of Creation* had caused a stir among the public and professional classes in England; and Patrick Matthew, whose 1831 book *On Naval Timber and Arboriculture* contained in its appendix a discussion of natural selection in the evolution of trees. Darwin knew Chambers's book well but had no knowledge of Matthew's book until the resemblance was pointed out to him several years after the publication of *On the Origin of Species*. For a useful summary of these and many other predecessors of Darwin, see J. L. Dagg, Natural Histories, "Natural selection before Darwin and Wallace," https://historiesofecology.blogspot.com/2013/08/predecessors-of-darwin-and-wallace.html (10 August 2014).

† Alfred Russel Wallace came to the idea of natural selection independently from Charles Darwin, according to Wallace himself, while he was in a reverie waiting for a malarial fever to break. Intriguingly, Wallace had come to the idea in a similar way to Darwin's, that is, through close exposure to variation in nature, and the influence of Thomas Malthus's 1798 book *An Essay on the Principle of Population*. An engaging account of Wallace and his relationship to Darwin can be found in Chapter 8 of Michael Shermer's 2001 book, *The Borderlands of Science: Where Sense Meets Nonsense* (Oxford University Press).

‡ I am not including Wallace in this list, who came to the idea of natural selection via his own epic journey through natural history. See Shermer (2001), cited above.

to develop his own synthetic theory for heredity, which he trotted out for the world to see in his 1868 book *The Variation of Animals and Plants Under Domestication.*[16]

———•————

Darwin's solution to the heredity problem was a mechanism he called *pangenesis.* His idea was that adaptation and heredity were melded through the agency of *gemmules,* tiny particles that all the body's cells supposedly shed over the course of their lives. Gemmules were like little identification nanochips, containing information about the cell that spawned them: muscle cells shed gemmules that were distinct from those shed by bone cells, and both would differ from gemmules derived from brain cells, and so forth. These myriad nanochips circulated throughout the body during an animal's life and eventually came to rest in the germinal tissues: the testes and ovaries. There, the gemmules formed a sort of genetic parliament that decided what manner of gamete the sperm or egg would be. If a particular somatic tissue—say, the muscles and bones of the neck of a hypothetical proto-giraffe living in a savanna—was used more, or grew disproportionate to other tissues during the animal's life, gem-mules from the neck tissues would be represented in higher propor-tion in the germinal tissues of savanna-dwelling proto-giraffes than they would be in, say, forest-dwelling proto-giraffes, where the lush leaves were closer to the ground and access to them was easier. When breeding time came, embryos of savanna-dwelling proto-giraffes would start life with more neck gemmules and would develop more neck—longer necks—than would the offspring of forest-dwelling proto-giraffes. Carry this on for many generations and two distinct lineages would emerge: one in savannas leading to the long-necked giraffe, and the other in forests leading to its shorter-necked relative, the okapi.

This familiar example of the giraffe just-so story* should look familiar because it can be found in nearly every biology textbook written since. Usually, the giraffe story is filed under Lamarckism, but in all fairness it should be filed under Darwinism, for Darwin's theory of pangenesis was a Lamarckian scheme for the heritability of acquired characteristics across generations. The principal difference is that Lamarck's *influence des circonstances*—the adaptive force that caused disused organs, like cave fish eyes, to disappear over generations—was transmuted by Darwin into the tiny particulate gemmules that accomplished the same thing. Pangenesis thus puts Darwin squarely into the camp of the much-derided Lamarck.[17]

It's worthwhile asking, therefore, why it is Lamarck and not Darwin who is tarred with the brush of the inheritance of acquired characteristics, the wrong-headed theory of inheritance that supposedly discredits Lamarck. Part of it may be due to style: where Lamarck built grandiose unworkable systems, Darwin painstakingly ground his way through evidence, the quirkier the better.

The famous story of Lord Morton's mare provides a telling example of Darwin's approach. The Scot George Douglas was the 16th Earl of Morton (therefore, Lord Morton). He was a well-regarded breeder of horses: among his projects was an attempt to domesticate the quagga, a South African subspecies of the Plains Zebra. Unlike the zebra, whose coat is striped all over, the quagga's stripes were limited to the neck and some markings along the legs. The quagga also had in common with the zebra the stiff-haired mane, which set it apart from the horse's mane, which is long and flowing. The quagga, sadly, is now extinct, the last living mare having died in 1870

---

* Rudyard Kipling's *Just So Stories* (1902) serve as the colorful inspiration for a criticism of so-called *ad hoc* adaptationism, in which some trait is presumed to be an adaptation because it exists as a trait. Kipling's stories are fanciful tales for how animals come to be shaped the way they are. The elephant's trunk, for example, is long because a crocodile tugged on it when the elephant came to the river to drink. In the case of evolutionary biology, *ad hoc* adaptationism is regarded, correctly, as a tautological fallacy.

at the London Zoo. But they were very much present in 1820 when Douglas undertook his effort to domesticate the animal.

Douglas could acquire only a male quagga for his project, so he decided to breed his acquisition with one of his chestnut mares. The offspring of that mating, as would be expected, showed many characteristics of a hybrid mating: it mostly looked like a horse but had striping along the legs, stiff hairs along the mane, and some subtle zebralike shaping of the face and mouth (Figure 5.4). But what came next was decidedly odd. Douglas then mated his chestnut mare with an Arabian stallion, which you would expect to produce a colt that blended the traits of two horse parents: the chestnut mare and the Arabian stallion. Instead, the colt showed remnants of the striping and stiff-haired mane of its quagga stepfather. To horse breeders, this was a quirky mystery, a bit of lore to ponder as part of the horse breeder's art. To Darwin, it underscored a deep lesson about heredity. How else could the story of Lord Morton's mare be explained unless there were some form of heritable particle—quagga gemmules, we might call them—that could impart "quagga-ness" onto the chestnut mare's future ova?

Figure 5.4
The offspring of Lord Morton's mare. *(left)* The quagga-horse hybrid. Note the striping along the neck and hind legs and the stiff, short hairs of the mane. *(right)* The subsequent offspring of Lord Morton's mare and an Arabian stallion. Note the striping of the hind legs and the stiff hairs of the mane. From Charles Hamilton Smith's 1841 *The Natural History of Horses*.

Add to this quirky little story twenty-six chapters of character-istically dense Darwinian prose, painstakingly bolstered by other examples drawn from hereditary diseases of the eye to the variegated leaves of plants to the effects of heat on the hides of South American cattle, and you come out with Darwin's pangenesis theory. Darwin came to the only conclusion he reasoned could explain these innu-merable examples: it had to be pangenesis, or something like it.

Today, we look upon Darwin's pangenesis adventure as a quaint anomaly, something akin to a revered ancestor's flirtations with a secret society that involved funny hats and odd rituals in secret rooms. *What could the old coot have been thinking?* we ask ourselves as we indulgently shake our heads. The joke is on us, though, because in proposing pangenesis, Darwin saw, more clearly than we typi-cally bother to do today, the essential truth that Lamarck also saw: that evolutionary and physiological adaptation somehow shared a common foundation. In case we are tempted to dismiss pangene-sis as a curious anomaly, it's noteworthy that Darwin published this idea nine years after the first publication of *On the Origin of Species* and so represents Darwin's mature thinking.

Pangenesis was not exactly well received by Darwin's contem-poraries, however. Although pangenesis provided a theory for the matter of "telegony" (literally reproduction at a distance, or the ten-dency of a parent to influence the traits of a lineage many genera-tions hence), as exemplified by Lord Morton's mare, the evidence for it was nevertheless mostly circumstantial. Even though Darwin had made the best circumstantial case he could make in *The Variation of Animals and Plants Under Domestication*, his case did not stand up well to experimental test.[18] For example, Darwin's own half-cousin, Francis Galton, took pangenesis to test by transfusing blood between rabbits of different coat colors to see whether the supposed blood-borne gemmules from rabbits of one color could affect the coat colors of subsequent generations of pups. Galton couldn't find any effect, leading him to reject pangenesis. This caused some hard feelings in the extended Darwin family because Galton had undertaken and published this rebuke without consulting his famous cousin.

So negative were the reactions to pangenesis that Darwin eventually likened his idea to being a "mad dream."[19] But he never really disavowed the idea, mostly because of all that pesky evidence he had compiled. There had to be some explanation for it all, and until there was a better one, pangenesis was the explanation that he had found convincing. As an aside, pangenesis seems lately to be making a comeback. We now know, for example, that DNA and cells from embryos can enter the body and organs of the mother and affect her subsequent offspring, just as Darwin's gemmules were supposed to have done.[20]

In telling this story, there is a broader point I wish to make, however. Darwin was driven to propose pangenesis because he recognized, again more than we do today, the essential tension that sits at the heart of the evolutionary idea: between adaptation, which implies the ability to change according to conditions, and heredity, which is the opposite of change, that is, the legacy of the past imposing itself on the future. We should take a moment to let that sink in. Lamarck, for example, was all about change—about forces that push organisms and their lineages in grand arcs of greater complexity and ever-improving fit to environments. Cuvier, for his part, was all about *resistance* to change, lineages and species held in place by the glue of the conditions for existence that bind all of an organism's many parts to one another—the unchanging legacy of one generation to the next. Today, we conflate evolutionism with Darwinism rather than Lamarckism or Cuvierism because Darwin was able to see, in ways that neither Lamarck nor Cuvier did, that a coherent theory of evolution must reconcile these contradictory ideas, that the tension between adaptation and heredity had to be resolved somehow.

As it turns out, Darwin was unable to resolve it either, so the essential dilemma remains and has shaped much of the subsequent history of evolutionary thought. Its practitioners would argue that we resolved the dilemma in about the mid-twentieth century, as Michael Ghiselin did in the thumbnail history with which we began this chapter. The implication is that we now have in hand the coherent theory of evolution that Darwin started but failed to complete, and that theory is thoroughly mechanist, materialist, and purposeless. But the story, as we shall see in the next few chapters, is more complicated than that, mixing in roughly equal measures of brilliant insight, unresolved contradiction, and bedazzling sleight of hand. In the end, the task of reconciling the tension between adaptation and heredity remains incomplete.

Before continuing our journey, I want to put Darwin and Bernard into some context, for each was, in his own way, reflective of the Hobson's choice of modern science. Let me present the Hobson's choice another way: when we study nature in a scientific way, should we approach it as Romantic idealists or as Enlightenment rationalists? Must we be one or the other? Cannot we be both?

We saw in Claude Bernard's life how he straddled those two worlds, mixing clear-eyed rationalism with Romanticism: homeostasis is, after all, a very idealist conception, very much a Romantic idea. Darwin, for his part, similarly straddled both worlds, one foot grounded solidly in evidence apprehensible by the senses but the other stuck in the essentially Romantic idea of adaptation, of the ineffable coherent organism. The two men's lives pose for us an uncomfortable question: would we today consider either Darwin or Bernard scientists? I suspect that the answer would be "no," given the difference between the mythic characters both are widely portrayed to be, and the way they actually were. Ultimately, both have suffered the same fate at our hands. Just as Bernard has had his Romantic roots cut out from under him as physiology cast its lot solely with mechanism, so too has Darwin seen the vitalist and Romantic core of his theory reamed out by his successors. The intellectual legacies of both men have been impoverished by our ministrations.

6

THE BARRIER THAT WASN'T

In the authoritative opinion of the indefatigable Ernst Mayr, August Weismann was the greatest evolutionary thinker of the nineteenth century, save Darwin himself (Figure 6.1).[1] Among other things, he helped launch twentieth-century biology on its all-consuming quest for the material nature of the gene. In so doing, he set biology's trajectory toward its twentieth-century materialist triumphs and, as it turns out, toward its looming crisis.

Today, Weismann's reputation is mostly pegged on his decisive purge of the dreaded taint of Lamarckism from evolutionary biology, where only pure Darwinism is allowed to rule. Because of Weismann, we now regard Lamarckism as a quaint philosophy with a ruffled collar that we need no longer take seriously. We keep it around mainly for nostalgia, because it might have inspired Darwin in his thinking. But once Darwin was done with it, we have permission to put it back on the bric-a-brac shelf with all the other mental clutter we keep around for sentimental reasons.

The real story, as you might expect, is quite different. In the post-*Origin* nineteenth century, Lamarckism ran rampant on the evolutionary stage and formed the main scientific opposition to the

Figure 6.1
August Weismann (1834-1914).

Darwinian idea. So serious was the Lamarckian opposition that by the end of the century, Darwinism was looking rather tattered, sinking into what has come to be known as the crisis of Darwinism.[2] In Darwinism's epic story, it was August Weismann who rode in on a white horse and saved the day, not through forcefulness of opinion or prestige of position, but with brilliantly clear reasoning and simple experiment. When Weismann was done with it, Lamarckism lay impaled and gasping on the Darwinian pickets, never to rise again. From that point on, evolutionary thinking could be neatly divided into two types: Darwinist and wrong.

That's the story we tell ourselves anyway. As usual, quite a bit of contrarian detail lurks within that comfortable myth, detail that we commonly regard as best forgotten. It is unnecessary to take Lamarckism seriously, we often tell students, because we have known since Weismann that Lamarckism is just plain wrong. Why delve into it? To do so would be like including a lesson on buggy whip technique in a driver's ed course: what could possibly be the point? There is a point, nevertheless.

August Weismann was born in Frankfurt into a solidly middle-class family. He was infatuated with butterfly and caterpillar collecting as a youth (inspired by his piano teacher), but his family's upwardly mobile aspirations drove him to pursue a career in medicine. He did not find medical practice much to his liking, and soon after completing his medical degree, he left for the more verdant fields of academic medicine. There, he began studying embryonic development and thinking about the new theory of cells that was then sweeping biology.* An eye disease eventually drove him away from the microscope and laboratory bench, but to such a man as Weismann, near-blindness was but a trifle—something that simply was not allowed to stand in his way. So he shifted his attention to thinking about the relationship between embryos and evolution and devising experiments to test his thoughts.

In the early post-*Origin* world, embryos were ground zero for the debate over Darwinism, because Lamarckism had found its surest foothold in embryonic development. To the embryologists, embryos were the parchment on which the history of evolution was written. New species, genera, phyla—all arose from subtle variations of embryonic development. New species were not so much new *things* as new *recipes* for how to build an organism. Imagine how a lean-to might differ from a pup tent or a tipi or a yurt. Each type of shelter differs in form—they are different *species* of shelter—but at root, they differ in the steps one follows to assemble poles, stakes, lines, and sheets into a shelter. So it is with embryos: new forms of

---

* The cell theory was enunciated in the early 1800s by Theodor Schwann, Matthias Schleiden, and Rudolf Virchow. Although the existence of cells had been known since the seventeenth century, cell theory rested upon two dogmas: that the cell was the fundamental unit of all living things, and that cells could arise only from other cells. Weismann's interest in cell theory was driven by technical advances in microscopy in the mid- to late nineteenth century that fueled a resurgence of work on cells, their structure, and how they worked.

organism come from how and when the embryo folds, grows, and pinches itself into new forms. Evolution, to the embryologists, was the historical record of how these basic building processes elaborated into new organisms: shuffle the steps, add a new step here, delay another step there, and you have a new organism. To those who looked upon evolution as the emergence of new forms of organisms, it was difficult to see how Darwin's model of new species arising by natural selection could possibly work. Where is the competition and struggle for existence in the protected and inexorable development of the embryo? Darwin himself had no good answer for this rhetorical question: in fact, his only answer was not Darwinian at all, but his Lamarckian model of pangenesis.

Weismann is supposed to have cut through all this by building a wall, what has come to be called the *Weismann barrier*. Weismann himself described it as the "segregation of the germ line," which requires some explanation, of course. Let's start with the "line" part. All embryonic development is a story of lineages, a family tree of sorts that represents lines of descent from progenitors (Figure 6.2). In the case of the embryo, the ultimate progenitor is the single fertilized egg, the zygote. The zygote divides, producing two descendants, which themselves divide, and *their* descendants divide, ad infinitum, producing the multitude of descendant cells that compose the mature organism. Each cell of the body can therefore trace a line of descent back to that single zygote, just as I can trace the Turner lineage in North America back to the unfortunate John Turner who sailed on the *Mayflower* to the New World.*

As for the "germ line" part, Weismann proposed that among the multitudinous lineages of cells within the developing embryo, two

---

* According to my family history, both John Turner and his son traveled to the New World from Leiden (via Plymouth) on the *Mayflower*. Both died within five weeks of their landfall at Plymouth Rock. The Turner family was established in the New World when John Turner's widow came in a later passage with her other children to Plymouth Colony.

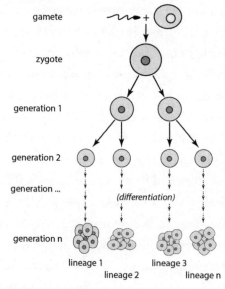

Figure 6.2
Lines of descent in the development of the embryo.

distinct categories of cell lineages stand out. On the one hand is the somatic line, or *soma*, which builds the body and all the physiological functions that come with it. On the other hand is the *germ line,* whose cells segregate themselves from the soma early in embryonic development and eventually become the gametes—the ova and sperm (Figure 6.3).

The soma and germ line are components of embryonic development, to be sure, but both also play into evolution in distinctive ways. The soma is the essential vehicle that nurtures and readies the gametes for reproduction, but the soma—so resplendent in our eyes—is merely the supporting actor for the real stars of the evolutionary show, the gametes. Once the gametes have been ushered through their role, the soma shuffles offstage and dies.

There is another important distinction between germ line and soma, and this defines the "segregation" part. To produce the mature organism, the somatic lineages must *differentiate,* that is to say, they must specialize to perform specialized functions. So, for example, one lineage may specialize to produce muscle cells, another will specialize differently to produce bone cells, other lineages will specialize as brain cells, skin cells, etc.* To Weismann, such differentiation could never be allowed to happen in the germ line, because that would compromise the fidelity of the soma across generations. The undifferentiated germ line was what ensured that mouse parents would produce mouse children and not, say, sparrow children. The germ line therefore had to be kept pristine, walled off from the tumultuous fecundity of its somatic siblings. The wall that keeps soma and germ line apart is Weismann's barrier (see Figure 6.3).

The Weismann barrier is important because it throws an entirely new wrinkle into the central paradox of Darwinism: the tension between adaptation (which requires change) and heredity (which opposes change). As we have seen, Darwin reconciled this tension with his theory of pangenesis, which allows physiological adaptation in the soma to affect the germ line through the gemmules. This is a very Lamarckian view, because both the organism and its lineage share a common language of adaptation, of the organism's "many little lives" negotiating their passage through time, mediated by the traffic in gemmules between the body's various parts, and across generations, as suggested by the stripy offspring of Lord Morton's mare (Chapter 5). Weismann's barrier drove a wedge through all that. Now, physiological adaptation in the soma could no longer affect the germ line. No longer could evolution be the outcome of a turbulent and raucous conversation between soma and germ line. Now, the germ cells were elevated to be the organism's cloistered royalty, never to be sullied by mingling with the grubby and grasping lumpen soma. Evolutionary change could come about only

---

* In the human body, for example, there are roughly two hundred distinct cell types, each representing a differentiated lineage from the zygote.

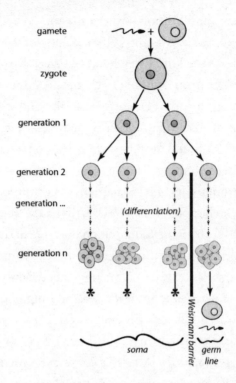

gamete

zygote

generation 1

generation 2

generation ...

(differentiation)

generation n

Weismann barrier

soma

*germ
line*

Figure 6.3
Segregation of the germ line from the soma by the Weismann barrier. The continuity of the germ plasm is ensured by the production of new gametes from the germ line. The somatic lineages die.

through changes in those pristine germ cells. The "many little lives" of the soma, for their part, were reduced to something akin to serf-dom. Like serfs, the soma's fleeting existence on the other side of the Weismann barrier mattered not a jot to the cloistered royalty within.

———————

Weismann didn't just *think* Lamarckism into oblivion: he was supposed to have killed it off with a brilliantly simple experiment. What happens, Weismann asked, to a lineage of animals if the soma is modified, generation upon generation? If the soma's experience informed the germ line, as Darwin and his Lamarckian rivals said it should, persistent modification of the soma should eventually show up in the lineage. So, Weismann took a colony of mice and amputated their tails. He bred these newly tailless mice with one another, amputated their offspring's tails, bred them and amputated *their* tails in turn, and so on, for five generations. In the end, the tails in the fifth generation of mice had not shortened, not even a whit.* The result was clear: continual modification of the soma had left no heritable imprint on the lineage. That left the germ line as the only possible avenue of inheritance, and therefore of evolutionary change. With his amputated mouse tails, Weismann had wrought a decisive experimental disproof of the Lamarckian doctrine of the inheritance of acquired characters. For this, Weismann's experiment has come to be touted as one of the greatest biology experiments of all time.[3]

Weismann's experiment, however, was nothing of the sort. First, the idea of a mutilation experiment to test Lamarckism was not original to him: such experiments had been common throughout the late nineteenth century. Many of these experiments were carried out by animal breeders, who were inspired by the experience of their counterparts in plant husbandry. Horticulturists often saw that grafts take on characteristics of their root stock, and animal breeders naturally wanted to know whether they could apply similar art to their own hybrid creations. These attempts were generally failures. Weismann was familiar with this dismal record, so he was very confident what the result of his mutilation experiment with mice would be.[4] Why do the experiment then? It's a fair question to ask. Weismann also was familiar with the numerous folktales circulating at the time that acquired characteristics could be heritable, and he found them all wanting. He cites a claim, for example, that a race

---

* The number of generations in this experiment differs widely in different accounts of it. Weismann himself said five (see "The Supposed Transmission of Mutilations" in Weismann's *Essays upon Heredity and Kindred Biological Problems* [1889]). Some accounts put the number of generations as high as twenty.

of tailless cats in Germany had arisen from a female whose tail had been lost when it "was said" to have been crushed under a wagon wheel. The problem there, of course, was the circumstantial and hearsay "evidence," never mind the unknown father of the tailless line. In another, better documented case of supposed acquired feline taillessness, from the Black Forest village of Waldkirch, the culprit turned out to be the pet Manx tomcat of the village parson's English wife, which, as tomcats do, had surreptitiously sired an extensive lineage of offspring with diminished tails.[5]

There is also a certain nagging . . . illogic . . . to Weismann's experiment, which seems quite at odds with Ernst Mayr's assessment of him as a brilliant evolutionary thinker. Let's start with the experiment's core assumption: that mutilation of the body is somehow akin to the normal process of adaptation that Lamarckians thought guided evolution. Who believes that? No one I know, nor were there many takers of the proposition in Weismann's day. Nor is the division of germ line from soma always clear-cut: sometimes the segregation occurs very early in embryonic life, sometimes later, sometimes not at all. Why make segregation of the germ line so important? Germ lines are also a peculiarity of animals, leaving vast swathes of the Earth's biota uncovered by Weismann's doctrine of the segregation of germ lines: you can't segregate something that doesn't exist.

Weismann knew these details very well—he was a superb naturalist and embryologist—so it seems a little strange that we are asked to believe he would pin so much on such a flimsy demonstration. If anything, Weismann's serial murine mutilation was not so much an experiment as it was a rhetorical flourish in an ongoing debate over something else entirely. By elevating Weismann's experiment into something it was not, we have lost sight of the point Weismann was trying to make. That point was not so much whether Lamarckism or pangenesis or the origins of lineages of tailless cats was true or not, but what the nature of heredity was. What Weismann did was to cast heredity into an entirely new framework, which colors how we think about it to this day.

———•———

Up to the time of *Origin's* publication in 1859, the best practical knowledge about heredity was to be found in the breeder's art, which was founded on the near universal belief that heredity was embodied in some sort of fluid "stuff": *germ plasm,* it was called. When a breeder did a cross, germ plasm from the parents was blended at conception to produce in their offspring a mixture of the parents' traits. This was how selective breeding was thought to work: improved lines of domesticated stock, animal or plant, were possible because the breeder, through crosses, could breed new lineages from parents with desired traits, mixing their germ plasms as a vintner blends various varietals to produce fine wines. Darwin knew this tradition well: selective breeding of pigeons, after all, was Chapter 1 in *Origin.*

Darwinism ran into early trouble because natural selection couldn't really work if inheritance was truly a matter of blending germ plasms. Lord Morton's mare, introduced in the last chapter, provides an instructive example. Lord Morton wished to develop a breed of striped horses that mixed zebra traits with horse traits. He approached this problem in the standard manner of the animal breeders of his day: he acquired a quagga jack (read, a conveyor of quagga germ plasm) to mate with his mare (similarly, a bearer of horse germ plasm). But, suppose Lord Morton had not had a quagga? Standard methods of artificial selection may have enabled him to pull off the trick without one. Horses sometimes show faint striping of their coats. Lord Morton could have sequestered these striped sports, mated them with one another, and kept doing so until he ended up with a lineage of horses with stripes. His plan would be founded on the widely held rationale that the trait of stripiness in the horse's germ plasm, so faint in the normal strain of horses, would be concentrated with each successive incestuous mating. He would have to have kept strict control over which stallion mated with which mare (or jack with jinny), because the whole thing could otherwise unravel. The striped-trait germ plasm, "concentrated" by Lord Morton in the bloodline he had so painstakingly built, would be "diluted" with every cross with a wild-type horse.

The problem for *natural* selection, seen by all, including Darwin himself, was the unlikelihood that such strict control of a bloodline could be sustained in nature. The problem is a statistical one. Imagine that a sport—a stallion with stripes—emerged in a population of wild horses (a striped mare would equally do). It's easy to imagine that this striped sport might enjoy better camouflage, may therefore be more likely to escape the notice of sharp-eyed predators, and consequently be more likely to survive and reproduce. Fine up to now; that's natural selection at work—but what then? What sort of mate would our lucky striped stallion be likely to find? Most probably, it would be a uniformly colored mare. Sports are, by definition, rare, making the "nonsports" or "wild types" abundant and more likely to be mating partners. Blending of their respective germ plasms would then produce offspring that, if they were striped at all, would be less distinctly so. If these dimly striped offspring managed to survive and mate, it would also likely be with other more abundant plain-colored horses, further dimming the striping in *their* offspring. With every generation, the trait that conferred the supposed selective advantage—stripiness—would be diluted more and more. In that day's parlance, the new trait would be "swamped." Natural selection—heritable adaptation—therefore could not work in a regime of blended inheritance.

For natural selection to work, the "stuff" of heredity therefore had to be something other than a fluid. We see the alternative expressed in Darwin's pangenesis theory: traits that were carried on hereditary *particles* of some sort. This solved the problem of swamping: heritable fluids could be diluted, but particles could not, and this meant that particles could be passed unchanged from generation to generation. Darwin's gemmules were only one aspect of particulate inheritance, however, because they carried only one kind of heritable memory, the "soft inheritance" of Lamarckian adaptation. Soft inheritance was fine for explaining adaptation, but a complete theory of particulate inheritance would also have to include particles of "hard inheritance" that ensured fidelity of the soma. It was hard inheritance that ensured that mouse mothers gave birth to mouse pups and not to sparrow chicks. This was the nineteenth-century resolution of that essential tension between adaptation (requiring change) and heredity (opposing change). Darwin's conception of gradual evolution of one species to another was based upon some interplay of the particles of soft inheritance—adaptability—with particles of hard inheritance—constancy. Darwin's mechanism—natural selection—could work if the conservative tendencies of hard inheritance could be nudged gradually by the ameliorating and adaptive tendencies wrought by soft inheritance. This is why Darwin needed some Lamarckian form of heredity for his theory to work.

Despite its logical necessity and Darwin's closely argued justification for it, his gimcrack pangenesis scheme—his "mad dream"—found little support among his contemporaries. Weismann himself described pangenesis as "rather distinct from reality," which is a tad unfair because no one, Weismann included, had any good idea of what heredity was or how it worked. This is why the post-*Origin* nineteenth century is such a confusing time in the history of evolutionary theory; it's difficult to understand and therefore difficult to teach. Who in that time, for example, could claim to be "Darwinian"? The self-proclaimed "Lamarckians" conceived of heredity in much the same way that Darwin did: wouldn't it be more fitting to call them Darwinians rather than Lamarckians? Meanwhile, self-proclaimed Darwinians like Weismann had cleaved Darwinism in two by wedging barriers between the interplay of adaptation and heredity that Darwin himself had thought so crucial. Should they have been calling themselves Darwinians? To be fair, they both were really Darwinian, the differences between them being a dispute over which of the twin pillars of Darwinism—adaptation and heredity—they thought important. The convenient labels of "Lamarckian" or "Darwinian" obscure this distinction and do little to clarify who was who and who thought what.

———•———

The theory of *orthogenesis* (literally "origin in a straight line") provides an interesting example of how all this played out in the turmoil of the post-*Origin* nineteenth century.[6] Orthogenesis was the focus of much of the "scientific" case against Darwinism at the time.

Like natural selection, orthogenesis was a model for gradual evolution, of the slow transmutation of species over time. Where Darwin proposed natural selection as the driver, however, orthogenesis was supposedly driven by a kind of "evolutionary inertia" that impelled lineages through time on a grand trajectory as if they were ships on a grand cruise. Since we are talking a lot about horses and other equids, let's illustrate the difference using the evolution of the horse family.

The ancestral horse was *Eohippus* (dawn horse), a forest-dwelling creature that lived about fifty million years ago. Dawn horses were small, about the size of a dog, with two toes on their feet. Horses (*Equus*), of course, are large single-toed hoofed animals, commonly inhabiting steppes and other grasslands. Let's look at two hypothetical family trees for the equids that plot body size versus time (Figures 6.4 and 6.5).

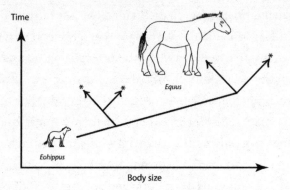

Figure 6.4

A Darwinian family tree of equids. When driven by natural selection of lineages, the trajectory from *Eohippus* to *Equus* sometimes regresses to smaller body size (left-pointing arrow), sometimes to larger body size (right-pointing arrow). The path from the small *Eohippus* to the larger *Equus* is not straight.

A Darwinian scheme of evolution would predict a highly branched family tree (see Figure 6.4). The overall trajectory would be from small *Eohippus* to large *Equus,* but along the way, various lines would branch off as intermediate species—*Mesohippus, Pliohippus,* to name two—all of which are now extinct. Sometimes these intermediate species would be smaller than their immediate ancestors, indicated by a left-pointing branch; sometimes, they would be larger, indicated by a right-pointing branch. The point is that the evolutionary trajectory from small *Eohippus* to large *Equus* was not direct. It would meander around, branch here and there, most branches becoming extinct, some carrying on to found new ancestors.

The orthogenetic scheme of equid evolution would have produced a much different family tree (see Figure 6.5). Now, the trajectory from small *Eohippus* to large *Equus* would have been quite direct, almost as if it were being impelled along a track. Natural selection could have played a role here, but it would have been minor, with little fringe branches here and there, nudging the trajectory this way or that, but having little effect on the lineage's grand journey through time from small to large.

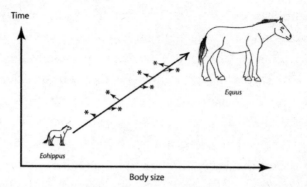

Figure 6.5
An orthogenetic family tree of the equids. The evolution of the equids from *Eohippus* to *Equus* is mostly a straight trajectory of increasing body size through time. Natural selection might nudge the trajectory one way or the other, but the drive to increasing body size dominates the growth of the tree.

This difference is something that can be tested with the fossil record. If Darwinian natural selection were driving the thing, branched family trees should be the norm; if it were orthogenesis, the fossil record should be dominated by the grand trajectory of continuously increasing body size. Alas, both sides could point to convincing evidence from the fossil record, which settled nothing.

Nevertheless, orthogenesis became one of the bludgeons wielded by anti-Darwinian evolutionists. Some wielding the bludgeon were strict materialists, bolstering their arguments with complicated theories of the nature of heredity. Others were "Neo-Lamarckian," who objected to the "Neo-Darwinists" shoving adaptation aside in favor of heredity. The Neo-Darwinists, for their part, hit right back. They painted the Neo-Lamarckians with the tar brush of retrograde vitalism motivated by nostalgia for a rosy Lamarckian past. The materialists were criticized by the Darwinians for overly speculative theories of heredity, again an unfair criticism because everyone was groping their respective ways forward equally in the dark. As I have said, this time is confusing and complicated.

Wilhelm Haacke (1855–1912), for example, had a strictly materialist explanation for orthogenesis purged of any form of vitalism or higher purpose. To Haacke, orthogenesis was brought about by the limited ways that structural elements within cells, which he called *gemmaria,* could combine to produce the different organs and cell types within the body. Just as a pair of dice produces a limited set of rolls, the combinations that gemmaria could form were also limited, and this meant that only a limited set of body forms was conceivable. The tendency of lineages to evolve along particular lines—the objective phenomenon of orthogenesis—was simply evidence that the combining rules of the gemmaria kept lineages within those lines of possibility. Orthogenesis, according to this conception, was no more remarkable than the tendency of a train to be guided by its tracks—and no more imbued with vital spirits or purpose than the train would be. To Haacke, orthogenesis was evolution reduced to a sort of physics, neither Lamarckian nor Darwinian.

Darwinians like Weismann were critical of Haacke's gemmaria because they admitted no role for adaptation: gemmaria are strictly particles of hard inheritance. This is why Darwin had to invent particles of soft inheritance—gemmules—to explain adaptation, despite having no evidence whatsoever of their existence or nature (and according to his half-cousin Francis Galton, evidence of their *non*existence). This tunnel vision is also why Darwinians, including Weismann, tended to be critical of orthogenesis generally. Darwin himself attributed the phenomenon of orthogenesis to artifact, an apparent pattern that would disappear as new discoveries could fill in the gaps in the incomplete fossil record.

There was another approach to orthogenesis in the post-*Origin* nineteenth century, one that tried to cut through all the flailing over particles of heredity, soft and hard. This approach was explicitly Lamarckian and was inspired by perhaps the most enduring example of orthogenesis to come out of the post-*Origin* period: Cope's Law.

Cope's Law is named for the most famous of the Neo-Lamarckian critics of Darwinism: the American paleontologist Edward Drinker Cope (1840–1897). Cope is best remembered today for his prodigious work in vertebrate paleontology, which thrust American science into world prominence when many regarded the phrase "American science" as an oxymoron. He was one of America's premier evolutionists at a time when that species was rare on the ground.

Cope came from a prominent Quaker family in Philadelphia and he lived most of his life there, the exception being a stint as professor of natural history at Haverford College, just outside of Philadelphia. He was part of the city's intellectual elite: founder of the American Philosophical Society; founder, long-time editor, and publisher of *The American Naturalist;* curator of vertebrates at the Academy of Natural Sciences; and professor of geology at the University of Pennsylvania. The prominent herpetological journal *Copeia,* still in publication, bears his name. Cope was also a colorful (and by many accounts, quite difficult) character (Figure 6.6).[7] He seemed to feed off confrontation, the most famous being his long-standing feud with Yale's Othniel Charles Marsh in the famous "bone wars," which opened the spectacular fossil beds of the American Midwest and elevated Cope to folkloric status.[8] He had a sad end, dying alone on a cot in his dusty house in Philadelphia, surrounded by crates of the fossils that he had spent a lifetime accumulating and for which he had not been able to find a home he deemed proper for their ultimate repose.

As with most folkloric characters, death did not stop the growth of Cope's colorful legend. To give one example, Cope had requested in his will that his brain and skeleton be donated to the American Anthropometric Society, which was dedicated to studying the brains of highly intelligent men. The Anthropometric Society duly accepted Cope's remains, but being concerned with brains, did not

Figure 6.6
Edward Drinker Cope (1840-1897).

know what to do with his bones. So, Cope's skeleton was turned over to Philadelphia's Wistar Institute, which had had a loose association with Cope, but which didn't know what to do with it either. Cope's will contained no instructions for such a contingency, so his bones were boxed up and stored in the Wistar Institute's basement.

There they sat until the 1960s, when Cope's skull emerged at the center of a bizarre dispute between the famed Darwin historian Loren Eiseley, Robert Bakker—the flamboyant paleontologist behind the hot-blooded dinosaurs controversies of the 1980s—and Wistar.[9] Eiseley had asked Wistar to borrow Cope's skeleton, ostensibly so that his skull could be used for a sculpture. Instead, Eiseley had made Cope's skeleton into a makeshift office shrine, the object of toasts for privileged visitors. In the meantime, Bakker felt that Cope would be better honored by his skeleton being offered as the holotype specimen for *Homo sapiens*.* So Bakker absconded with the

---

* Bakker's quest went nowhere. An apocryphal story emerged that Cope had made the holotype request himself in his will, but that his skeleton had been rejected as the holotype because it was deformed by syphilis. None of this is true. *Homo sapiens* has a perfectly good holotype (Carl Linnaeus himself), Cope had made no such request in his will, and there is no evidence that Cope suffered from syphilis, despite a well-established reputation as a womanizer. For a fuller history of Cope's life, see J. P. Davidson's 1997 biography *The Bone Sharp: The Life of Edward Drinker Cope* (Philadelphia, Academy of Natural Sciences of Philadelphia). For a fuller account of the bizarre story of Cope's skeleton and skull, see D. R. Wallace's 2000 history of the rivalry between Cope and Marsh *The Bonehunters' Revenge: Dinosaurs, Greed, and the Greatest Scientific Feud of the Gilded Age* (Houghton Mifflin).

skull, which somehow turned up in the possession of a photographer and documentary filmmaker, Louie Psihoyos, who constructed a velvet-lined reliquary and took the skull around the country so that he could photograph herpetologists and paleontologists with the precious relic. Eventually, the skull was recovered, and now it sits at the University of Pennsylvania Museum of Archaeology and Anthropology.*

But, back to orthogenesis. Cope's eponymous law asserts that a lineage, once set apart enough to be distinctive, tends to evolve toward ever larger body sizes. Whenever a pattern like this crops up, two questions naturally arise: first, is it a real pattern?; and second, what's driving it? With respect to the first, Cope was no theory-spinner like Lamarck; he justified his law from analysis of empirical patterns of body size he found in the vertebrate fossil record, which, thanks largely to his own efforts, was much more complete and informative than it had been in Darwin's day. Since Cope, paleontologists have found many quibbles with his law: is it general enough to be a "law" or a "rule," or a "generally prevailing condition"?;[10] does it indicate evolution *toward* large body size or *away* from small body size?, etc.[11] Suffice it to say that Cope's Rule (let us call it that as a nod to appropriate modesty) has stood up pretty well.[12]

This brings us to the second, and more interesting, question of what drives it. To many, this is where Cope, and orthogenesis generally, begins to jump the rails of scientific respectability. Cope explained his own rule by invoking an inherent striving of living systems toward large body sizes, driven by an internal force he called *bathmism*. This idea has been likened to Lamarck's complexifying force, which has given Darwinians the ammunition to dismiss Cope as a misguided Lamarckian. They are being quite unfair in doing so, however, and it's worth delving into the reason why.

---

* The story of Cope's skull and Louie Psihoyos is told vividly but tendentiously and unfairly to Cope by Mark Bowden in his 1994 story "On the Trail of a Wayward Skull: Decades After Death, a Scientist and His Racist Ego Live On" (*The Philadelphia Inquirer*, October 6 1994).

Bathmism was, for its time, a sophisticated thermodynamic theory of evolution. According to Cope, evolution is driven by a kind of energy, what he called growth-energy or bathmic-energy. This was the first source of trouble for Cope. Where the mainstream of evolutionism was turning its gaze toward heredity, Cope was arguing that focusing on heredity to the exclusion of the forceful drivers of evolution was to miss a very big point about evolution and life. This contrarian approach, so typical of Cope's personality, made him a forceful critic of "Darwinism" as it was coming to be defined by Weismann and the other post-Darwin Darwinists. He succinctly posed the problem in this way: "The evolutionists [Darwinists] attempt to explain design in structure through the operation of the 'survival of the fittest' . . . It is justly argued against this reasoning that it attempts no explanation of the origin of such structures."[13] We can paraphrase this critique with what I call Cope's question: never mind the origin of species, what is the origin of *fitness*? How, precisely, do living things come to be fit in the first place?

Taken alone, Cope's question sounds cryptic, but it makes better sense in the context of his theory of bathmism properly understood. Cope was a vitalist, in the same sense that Claude Bernard was: he regarded life as a unique phenomenon in the universe, which demanded a unique mode of understanding. Where Bernard focused on homeostasis as life's distinct quality, Cope sought to frame the question in terms of energy, which led him to a surprisingly advanced theory of evolution—and of orthogenesis. He argued that there was a competition between two forms of work in the universe. *Anagenesis,* or order-producing work, would fuel growth and complexity. Anagenesis was balanced against work that drove matter toward disorder and death—that is, *katagenesis,* or degradation to entropy and equilibrium. Cope regarded living systems as anagenesis prevailing over katagenesis. In nonliving systems, the prevalence was reversed: katagenesis prevailed over anagenesis. Bathmism, or growth-energy, was what tilted the prevalence in living systems toward anagenesis.

Strip away the funny neologisms, and it is clear that Cope would have been quite at home with the parable of the cauliflower and cumulus cloud introduced in the Preface. No one pays attention to Cope these days, though, because Cope's theory marks him as both a Lamarckian and a vitalist: two strikes in the modern Darwinian mind. Strike three came because Cope saw no reason to draw a distinction between soma and germ line as Weismann had done. If the principle of bathmism applied to individual creatures, as Cope had asserted it does, on what basis can we say that it could not also apply to lineages? Both are expressions of life, and if bathmism helps explain one aspect, why should it not explain all aspects? In this, Cope was rowing against the tide, as Cope always did. Whether he was right or wrong in doing so depends, of course, on the strength of his critics' arguments.

Which brings us back to Weismann's barrier.

<hr />

We can chuckle at Cope's quaint terminology and his flirtation with Lamarckism and vitalism, but the fact is that the Darwinists in Cope's time had no better answer. But wait, you might say, didn't Weismann put all that to rest with his impenetrable barrier? As it turns out, the Weismann barrier was not something that was demonstrated with beautiful experiments and empirical proof. Rather, it was a logical necessity that followed from Weismann's own totally erroneous theory of embryonic development.

In the developing embryo, a kind of evolution is at work.* The question that consumed Weismann was how the embryo's diverse

---

* Before Darwin appropriated the word "evolution" to describe the change of species through time, it was a term of art in embryonic development. There, the word described how the complex embryonic form emerges (literally *unfolds*) from the single cell of the zygote. Evolution in the Darwinian sense was often called "transformation" or "transmutation of species."

tissues and cells could arise from the single zygote: *differentiation,* in a word. Weismann was convinced that differentiation was governed by a kind of hereditary particle, which his observations had located as residing in the cell's nucleus. The zygote's nucleus thus contained within it the hereditary particles that could determine the fates of the zygote's innumerable descendants. The problem Weismann grappled with was how the same set of hereditary particles in the zygote could produce such diverse outcomes in its descendants.

Weismann had a curious name for these particles—he called them *ids.*[*] Innumerable ids were contained within the zygote's nucleus, constituted from ids contributed equally by the mother and father, and a cell's characteristics were determined by its collection of ids. There were muscle ids, brain ids, skin ids, and so forth, all contained within the zygote's nucleus, ready to act. Ids could also be either active or dormant and could be wakened by various undefined forces operating within the cells (vital forces seem to creep in everywhere).

So far, Weismann's conception of the id is looking a lot like our modern conception of the gene, and in fact Weismann's id inspired the formal definition of the gene that would come at the end of the nineteenth century.[†] Where Weismann went wrong was his conception of how the innumerable ids in the zygote came to determine the many different cell types in the embryo.

---

[*] Sigmund Freud actually looked to Weismann's theories of embryonic development as an analogy for his own model of the psyche. "Id" in Freud's psychology refers to the instinctual drives of the psyche, which are tempered by the ego and superego. In the case of Weismann's use of the term, it derived from an earlier conception of the hard hereditary material by the Swiss botanist Carl Nägeli (1817–1891), which he named *idioplasm.* Weismann's use characterized the idioplasm not as the hereditary fluid conceived by Nägeli, but as a collection of determinist particles—ids.

[†] By the Dutch botanist Hugo de Vries (1848–1935), who originally named the hereditary particle *pangenes,* a reference to his theory of *intracellular pangenesis,* which looked to hereditary particles that carried different traits. The term "pangene" was later shortened to our modern term, "gene." De Vries, along with Erich von Tschermak and Carl Correns, rediscovered Gregor Mendel's work on genetics, which had lain in obscurity for decades. Correns was a student of Nägeli, and it is known that Mendel had written to Nägeli about his work, so the concept of the pangene is probably traceable through Correns.

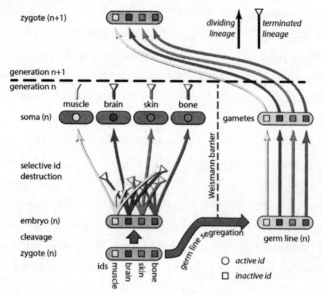

Figure 6.7

Weismann's conception of embryonic differentiation and development. Heredity was carried on particles that Weismann called *ids*. These were present in active forms (circles) or inactive forms (squares). Differentiation involved the selective destruction of ids, leaving a subset of ids that defined the differentiated cell. Reproduction involved the transmission of form from one generation (n) to the next (n+1), which meant transmitting a complete collection of ids.

Weismann's conception is a little complicated but easy to grasp if we follow what happens to four hypothetical ids: for muscle, brain, skin, and bone. At conception, all ids are present in inactive form in the zygote (indicated by the square symbols in Figure 6.7), and these are transmitted in their inactive forms during the zygote's first few cell divisions, what is known as the *cleavage stage*. During cleavage, the zygote rapidly divides several times, copying its collection of inactive ids into several cells as it does, including one set that is to be segregated into the germ line. The Weismann barrier starts at this point.

What follows is growth of the embryo and differentiation of the various lineages of the cellular descendants of the zygote. As these cells continue to divide, they pass on their collection of ids. As the various lineages differentiate, ids become active (indicated by the circular symbols in Figure 6.7). Differentiation is possible, though, only if one type of id is activated in a lineage. Muscle cells, for example, are muscle cells because only muscle ids are activated within them. Weismann thought that various ids were selectively destroyed through the progress of a cell lineage toward differentiation and specialization, so that only one type of id would remain in the cells of a differentiated tissue. In muscle tissue, for example, all the ids handed down from the zygote *except* for muscle ids would be eliminated from the cellular lineage (indicated by the triangular symbols in Figure 6.7). The same would be true for brain tissue: all ids except brain ids would be eliminated, and so forth for all the differentiated cell types. The soma comprises all these differentiated lineages, and all these lineages eventually die.

Weismann came to this model not because he had any positive evidence for it, but because he had a hunch: he thought it absurd that the mature differentiated cells of the soma would carry all the zygote's ids. As Weismann put it:

> It is highly improbable that all the determinants in the id of germ-plasm are carried along through all the idic stages of the ontogeny [i.e., those stages of embryonic development where differentiation is occurring]. . . . Why should Nature, who always manages with economy, indulge in the luxury of providing all the cells of the body with the whole of the determinants of the germ-plasm if a single kind of them is sufficient?[14]

As it turns out, Weismann was wrong in his hunch. If we equate the id with the gene, as came to be the case, we now know that every cell in the body carries the same genes as the zygote, indulging in the very luxury that Weismann thought absurd. Differentiation comes from the different tissues *expressing* different suites of that complete collection of genes and silencing others—muscle cells express a different suite of genes than brain cells do—but all cells contain the same collection of genes as the zygote. In short, there is no selective destruction of ids during embryonic differentiation, as Weismann thought had to be the case.

Nevertheless, Weismann's assumption was enormously consequential, because the *logic* of his selective destruction of ids leads directly to the twin doctrines of the segregation of the germ line and the Weismann barrier. The logical chain is quite simple. None of the differentiated cells in the soma could be suitable candidates for the soma's reproduction, because none of the somatic cells carried within them contain the complete collection of ids that would be required for successful reproduction. The consequence of differentiation therefore is ultimate irrelevance and death of the soma. Successful reproduction, in contrast, could only come from a set of cells set aside early in embryonic life that retained a complete set of the organism's ids: the segregated germ line. Furthermore, these privileged cells could not be permitted to participate in the normal life of the soma, because allowing them to do so would risk breaking up the complete set of ids that would be necessary for the organism's successful reproduction. They therefore had to be segregated from the soma: the logic of Weismann's model of development demanded it. Successful reproduction could come only from gametes passing on a complete set of ids to the next generation. Evolution *within* the soma—what we now call physiological adaptation or phenotypic plasticity—could occur, but this could in no way influence the ids in the segregated germ line, because that would imperil the integrity of the complete collection of ids held there. Evolution of the *lineage,* in contrast, could come about only from modification of the ids of the germ line.

It was on this flimsy foundation that the wedge was driven that would cleave modern biology to the present day—the wedge between soft and hard inheritance, between physiological and evolutionary adaptation, between living body and the crystalline purity of gametes, between vital life and its clockwork imitation. The irony, of course, is that the Weismann barrier was erected not upon the foundation of a trivial experiment with mutilated mice, but on August Weismann's deep knowledge of embryology, his subtle thoughts on the relationship between embryology and evolution, and his formidable logic. That logic turns out to have been wrong, and this means that the Weismann barrier—that scourge of Lamarckism, that foundation of modern evolution—is the barrier that wasn't.

Weismann also set in motion the epistemic closure of modern Darwinism. Before Weismann, evolutionary thought had been concerned with the interplay between adaptation and heredity, between change and stasis, between soft and hard inheritance—Darwinism's central dilemma, in other words. That is a very difficult dilemma to unravel, and much of the confusing turmoil of post-Darwinian evolutionism—the crisis of Darwinism—is testimony both to the complexity of the problem and to the richness of evolutionary thought at that time. After Weismann, Darwinism became a theory concerned solely with heredity, indeed with only one form of heredity: hard inheritance. Adaptation, the organism, all the messy complications that went along with the interplay of adaptation with heredity, became secondary, shoved to the back of everyone's minds.

At this point, I want to introduce a word of caution, just to set the stage for what is to follow. I hope I have persuaded you that evolutionary thought in the late nineteenth century was very rich. One of the casualties of the epistemic closure of modern Darwinism has been the appreciation of just how intellectually rich that era was. The temptation is to conclude that modern Darwinism therefore has become intellectually impoverished by comparison. That would not be correct, because an epistemically closed world can also be incredibly rich. With epistemic closure, the boundaries of the playing field change, which has little to do with the brilliance of the play. LeBron James will play brilliant basketball whether he is playing full-court or half-court: he will just play brilliantly within different bounds. That is largely the case for the post-Weismann epistemic closure of evolutionism. There is much brilliance in there to appreciate, but the appreciation must be tempered by the realization that we are now watching half-court basketball.

7

THE REVERSE PINOCCHIO

In the Dutch city of Delft, behind the Nieuwe Kerk, stands a statue, *The Milkmaid* (Het Melkmeisje), Wim T. Schippers's stucco-on-concrete representation of Johannes Vermeer's famous painting of the same name (Figure 7.1). The statue and painting are each beautiful in their own ways. They both capture the milkmaid's peasant solidity and her firm grounding (literally, in the case of the statue) in the everyday life of seventeenth-century Holland. Yet Vermeer's painting seems somehow *alive* in a way that Schippers's sculpture does not quite capture. Vermeer's milkmaid bends her head down in rapt attention to her task, drawing our attention in as she takes care to not spill a single drop of milk as she pours. The painting so glows with vitality that the milkmaid seems about to step out of the canvas, off to do the next chore of her busy day. In contrast, Schippers's milkmaid seems frozen in place, as if she was Lot's unfortunate wife who looked up at Sodom at precisely the wrong moment. She stares off, mouth agog, through only hints of eyes, to some distant scene that has caught her attention.

Figure 7.1
The Milkmaid in stone (*left:* sculpture by Wim T. Schippers, 1975) and oil on canvas (*right:* Johannes Vermeer, ca. 1660).

This creative tension between the living and nonliving, between painting and stone, between the vital and the inanimate, permeated much of the development of evolutionary thought through the late nineteenth and early twentieth centuries. At stake was nothing less than the soul of biology: whether we would study life as a unique phenomenon to be understood by its own rules, or whether we would flatten life under the rules that govern the rest of the material universe. By the 1940s, the question was largely settled, for most, in favor of the latter. For those who had held out that life could be understood as something vital and unique, well—to lift the title of an essay by Tom Wolfe—sorry, but your soul just died.*

For those who had decided to embrace biology's brave new world, the death of biology's soul was like the death of a beloved pet in its dotage: lamented, an occasion for sad nostalgia perhaps, but a blessing in the end. Yet in that death were sown the seeds of biology's impending crisis—a crisis that began building with the crisis of Darwinism, which, contrary to the received wisdom, was not settled at the turn of the twentieth century. Nor has it been settled by biology's twentieth-century triumphant march to materialism. We may have been able to disguise it, but in fact, we are neck deep in the crisis still. We are neck deep in it because the crisis lies in the still-standing irresolution of Darwinism's central paradox: the tension between adaptation and heritable memory.

In the post-*Origin* nineteenth century, the crisis largely played out in the interplay between hard inheritance and soft inheritance. Hard inheritance was the boot to the flank that drove life ahead under its pitiless command, while in soft inheritance reposed the hope that life somehow had control over itself after all. In the hopeful shelter of soft inheritance stood the Lamarckian theory of *vital* adaptation, that a life's experience mattered across the generations as well as within them—a hope that was reflected in Darwin's pangenesis idea.

This hope reverberated through biology into the first decades of the twentieth century, but it was fated ultimately to be crushed. The beginning of the end can be pegged to August Weismann's decision to cast biology's lot completely with hard inheritance, which set

---

* Wolfe's 2003 essay ("Sorry, but Your Soul Just Died," OrthodoxyToday.org, www.orthodoxytoday .org/articles/Wolfe-Sorry-But-Your-Soul-Just-Died.php) refers to developments in neuroscience, in particular, functional imaging techniques like fMRI that seemed to be revealing the mechanistic correlates of consciousness. The motivation driving this late-twentieth-century development was drawn from the same metaphysical well that marginalized the vitalist idea, which had begun at the century's beginning—that life is a mechanism, to be understood strictly as mechanism. Wolfe expanded on this theme in his 2004 novel *I Am Charlotte Simmons* (Farrar, Straus and Giroux).

Figure 7.2
Thomas Hunt Morgan in 1920.

the scene for ushering adaptation off the stage, leaving only hard inheritance to command our thoughts. The final coup was delivered some thirty years later by another giant figure of twentieth-century biology, Thomas Hunt Morgan (1866–1945; Figure 7.2). Ironically, Morgan himself thought at the time he had killed off not only Darwinism for good, but Lamarckism as well. He was premature: as we shall see, the Darwinian idea was revived just a few years later by three brilliant thinkers who pulled off one of the greatest scientific achievements of twentieth-century science: the so-called genetical theory of natural selection. But as we shall also see, this did not resolve the crisis, it only deepened it.

———— • ————

Let us turn to Morgan first. Thomas Hunt Morgan is best remembered for his pioneering work on the genetics of the fruit fly, *Drosophila melanogaster,* literally the black-bellied fruit lover. Whenever biology students anywhere in the world are taught genetics, whether they are in high school or college, they take a vicarious stroll into

Morgan's famous "Fly Room" at Columbia University.* This is usually an exercise in tedium for students—many undergraduate science laboratory classes seem designed to maximize drudgery—but with some imagination, students can conjure the atmosphere of the Fly Room in their minds: its insistent smell of overripe bananas, the multitude of little bottles filled with breeding fruit flies, the maggots writhing through the nutritious mush, and the laborious sorting and counting of tiny fly offspring under the microscope. What is not often captured in the tedium of the classroom genetics laboratory is the heady intellectual atmosphere that must have pervaded the Fly Room, for also hatched in this tiny space were many of the most important minds of twentieth-century biology. One cannot read the accounts and reminiscences of the Fly Room alumni without seeing in their reflections the endearingly tatty *beau idéal* of the scientific life—egalitarian, rational, improvisational, nimble, experimental, interactive.†

Morgan is important to our story because he cemented into place Weismann's radical division of physiological and evolutionary adaptation. What is ironic is that Morgan was a harsh critic of Weismann's scientific thought. To be fair, Morgan was a harsh critic of the scientific thought of nearly everyone besides himself, even of Charles Darwin. This contradiction is essential to understanding many of the tangled roots of modern Darwinism, for with Morgan came full epistemic closure on the matter. Once the evolutionist episteme closed, no role for soft inheritance could even be conceivable, rendering it effectively invisible. With the closure began the ultimate alienation of the science of life from life itself.

---

* Morgan's "Fly Room" was active from 1911 to 1928. For an engaging depiction of the Fly Room, see Alexis Gambis's dramatization *The Fly Room* (www.theflyroom.com).

† Morgan was lured from Columbia to Caltech in Pasadena in 1928, where much of the work that led to his being awarded the 1933 Nobel Prize was carried out. See Judith Goodstein's engaging history of Morgan, his work, and his students, "The Thomas Hunt Morgan Era in Biology," *Caltech Engineering and Science*, Summer 1991, 12–23.

Morgan's basic philosophy of science was positivist, which means that he believed that valid knowledge could only come from experimental demonstration and test. All else was irritating "speculation" and intellectually suspect, and Morgan found many reasons to be irritated with the biology of his time. Weismann's ids were the particular source of Morgan's irritation with him, and the inchoate strivings proposed by the Lamarckians (and to a large extent, the pre-Weismann Darwinians) were another. Outside Morgan's positivist world, natural history was mere storytelling, taxonomy was stamp collecting, and paleontology was infested with entrenched Lamarckians like the impossible Cope.

I hasten to add that I do not wish to portray Morgan as a curmudgeon, for by most accounts, he was anything but.* He sprang from aristocratic roots in his native Kentucky, and he grew up loving nature and natural history. Intellectually, he came of age studying under the famous William K. Brooks of Johns Hopkins University, one of the most prominent American embryologists of his day and, in the tradition of the day, a thoroughgoing Lamarckian. Many of Morgan's sterling qualities as a scientist and person—his collegiality, devotion to a problem, love of nature as it was, and genuine affection for students—found encouragement and endorsement as he circled in Brooks's orbit. Morgan's doctoral work on sea spiders reflected the prevailing scientific tradition that Brooks had inherited from his own mentor, Louis Agassiz: descriptive, empirical, and aimed at using close description of embryological development to uncover sometimes counterintuitive evolutionary relationships. Morgan (along with many in his scientific cohort) eventually broke decisively from

---

* Morgan came from a stock of old Southern planter aristocracy. His paternal uncle, John Hunt Morgan, was a renowned Confederate general known for his bold and mercurial battlefield tactics. His great-grandfather, John Wesley Hunt, was one of the early migrants to venture past the Allegheny Mountains into Kentucky, eventually to become a wealthy merchant, planter, and horse breeder. His maternal great-grandfather was Francis Scott Key, composer of "The Star-Spangled Banner."

this tradition. The reasons for the break are worth exploring.

Since the Enlightenment had dawned two centuries previously, scientists had been putting themselves forward as keepers of an alternative and superior source of meaning regarding the natural world. For biology, meaning had come from the uneasy tension between the positivist ideal—experimental, empirical, rational— and the vitalist tradition that looked for meaning in life's unique qualities—perception, intentionality, purposefulness. In the nineteenth century, biology's creative vigor stemmed largely from this tension. Some, like Darwin, or Bernard, managed the tension brilliantly. It is no coincidence that the post-*Origin* nineteenth century was a time of remarkable intellectual ferment.

Friedrich Nietzsche and the rise of German nihilism threw a very large spanner into the works of these pretensions, however. High on the list for elimination was any residue of vitalist thought. As the vitalist wells began to dry up, biology was left thirsting to find new sources of meaning, and the readiest succor was to be found in those sciences, such as physics and chemistry, that were explicitly and exquisitely materialist. Ernst Mayr's "physics envy" (see Chapter 4) was the symptom of the biologists' metaphysical thirst. Slaking that thirst came with a price, because biologists now had to make the Hobson's choice: drink from the well and become a "real" scientist, but abandon your soul, or keep your soul and be cast into the thirsty wilderness. It was no longer an option to straddle the two worlds, as Bernard and Darwin had done.[*]

When Morgan was beginning his professional career around the turn of the twentieth century, the nihilists' monkey wrench was still grinding away in the cogs, with biology being torn between the

---

[*] It is a schism that continues to this day, in fact. A recent example has been the controversy over sociobiological principles in anthropology, depicted vividly in Napoleon Chagnon's 2013 memoir *Noble Savages: My Life Among Two Dangerous Tribes—the Yanomamö and the Anthropologists* (Simon & Schuster).

positivists on the one hand and the scientific vitalists on the other. On the vitalists' side were Cope, Brooks, Agassiz, and the rest of the Lamarckian swarm, looking at their erstwhile positivist compatriots from across a widening and eventually unbridgeable chasm. To Morgan, the future seemed to be with the positivists, and so that is where he cast his lot.

He was a zealous convert. The tipping point for him probably came just after he finished his doctoral work with Brooks in 1890. From Johns Hopkins, Morgan undertook a year's stint at the Stazione Zoologica in Naples, where he came to know Hans Driesch, who was probably Germany's premier embryologist at the time. Driesch's approach to embryonic development was much different from Brooks's. Embryologists like Brooks had sought to describe, in exhaustive detail, every movement of cells and every contortion of an embryo, in the hope that knowledge would emerge from the embryo, like Venus emerging from the sea (*Venus anadyomene*)*— fully formed, voluptuous, and beautiful. In contrast, Driesch and his cohorts sought knowledge by poking embryos with e. e. cummings's "naughty thumb of science,"† mucking in and doing experiments with them: what would happen if *this* cell were taken from one location in an embryo and put somewhere else, for example?

With this approach, Driesch had already knocked one prominent theory of embryonic inheritance into the bucket. That theory, the "mosaic" theory of differentiation, promulgated by the formidable Wilhelm Roux, proposed that embryos differentiated by divvying up the zygote's heritable stuff as it divided into different cells. Roux had suggested that embryonic differentiation was like dealing out

---

* *Venus anadyomene* was the classical artistic motif of Venus (Aphrodite) emerging from the waves on an oyster shell as a fully grown virgin, made most famous by Sandro Botticelli's *The Birth of Venus* (*Nascita di Venere*, 1486).

† From e. e. cummings's beautiful poem "o sweet spontaneous": "o sweet spontaneous / earth how often . . . / , has the naughty thumb / of science prodded / thy / beauty . . . ," published originally in 1920 in *The Dial* and later republished in cummings's first collection of poetry, *Tulips and Chimneys* (1923).

playing cards from a stacked deck. In the gamete, the incipient differentiated tissues were predetermined by packets of "seeds" distributed throughout the egg: brain cells, muscle cells, and so forth.
These "seeds" were then parceled out into different cells, the *blastomeres,* during the initial divisions of the fertilized zygote. One
blastomere would have "brain cell seeds" in it; blastomeres fated to
become muscle cells would have "muscle cell seeds" in them; and so
forth. This conception will be familiar: Roux's mosaic theory was
the close cousin of August Weismann's theory of cellular ids.

To test Roux's theory, Driesch took embryos of sea urchins in
their very early stages of development, when the zygote had divided
into eight blastomeres or so, and separated the blastomeres from one
another. By the logic of the mosaic theory, the blastomeres should
have already divided the zygote's primordial hereditary loot among
themselves. Once separated, each blastomere, if it went on to develop
at all, therefore should have developed into an incomplete part of
an embryo. The blastomere that had received mostly "brain seeds"
should develop mostly into brain tissue, and so forth.

But Driesch's result was completely contrary to what the mosaic
theory predicted. Each of the separated blastomeres developed into
complete embryos, indicating that there had been no divvying up of
heritable seeds, as Roux had proposed. Rather, each blastomere had
received a full complement of the same heritable stuff the zygote
had. And that, in one fell swoop, was the end of Roux's mosaic
theory of differentiation: the "naughty thumb" of positivism had
prevailed, and dramatically so. To an adventuresome soul like the
young Morgan, it's hard to imagine a greater temptation to join the
team.

Driesch was an exemplar of a larger, and mostly European,
school of experimental embryology that at the time was engaged
with remarkably sophisticated studies of cells and nuclei in
embryos. These biologists had a larger agenda: embryonic differ-

entiation seemed to turn on the nature of the hereditary "stuff," and this made embryos the ideal test bed for sorting out what this nature was.* By this time, the old germ plasm idea had dried to dust and blown away, leaving the particulate theory of inheritance ascendant. Adding to the triumph was the rediscovery of Gregor Mendel's studies on particulate inheritance of peas, finally brought to life, like Rip van Winkle, from a thirty-five-year slumber. The quality of microscopes, meanwhile, had become so good that microscopists could now locate particular things in the cell. These included objects that seemed to have something to do with heredity, and it was looking increasingly likely that these particles of inheritance, of hard inheritance at least, were located in the cell's nucleus. When Morgan came onto the scene, the hunter was closing in on the prey: the "atom of heredity" itself. Embryos would be the arena where the understanding of the physical basis of heredity would be won.

All was not well, though, because some of the microscopical results seemed at odds with how Mendel's "atoms of heredity," if that's what heredity was, should behave. Resolving these contradictions grabbed the young Morgan by the ears, setting him on his scientific life's mission: to discover the physical basis of the gene. To do so, he would abandon embryology, along with Brooks, Agassiz, Driesch, and the rest, and take up the problem of inheritance. The fruit fly seemed to have a comparatively simple pattern of inheritance, and so this is how *Drosophila melanogaster* came to be the horse that Morgan would ride on his quest.

To fully understand Morgan's true greatness as a biologist, one

---

* In another of the strange ironies that permeate this period of evolutionary thought, Driesch himself became a leading Neovitalist, proposing that the tendency of blastomeres to become whole embryos was evidence of a determining tendency within cells that he called *entelechy*, after Aristotle's conception of the same name of an inherent tendency to form. Driesch went on to occupy a chair in natural theology at the University of Aberdeen and various chairs in philosophy at Cologne University and later at Leipzig University.

needs some details concerning what the contradiction between the microscopists and Mendel was all about. But don't worry, gentle reader, that is the province of the undergraduate biology course, and not something I wish to inflict on you. It is important to know, however, that by the time Morgan had finished his quest, Mendel's "atoms of heredity," which had existed only in statistical imagination, had become real, located on those strange bodies in the nucleus called *chromosomes,* even to the point of being mapped onto the chromosome like the stops on a subway line. Before Morgan, there was only a vision of the gene; after Morgan, there was a true science of the gene. I'm afraid you're just going to have to take my word on that—and the word of the 1933 Nobel Prize committee.

———————•———————

Morgan's brilliant vindication of Mendelian genetics also challenged the Darwinian idea to its core, and this is where we cannot avoid delving a little into details. On the one hand, the reality of the Mendelian gene that Morgan demonstrated bolstered Weismann's concept of particulate inheritance: Weismann's id was now unmasked as the Mendelian gene. On the other hand, Mendelian inheritance predicted patterns of variation that would be the *opposite* of what a Darwinian model for evolution supposed.

Darwinian evolution supposes that slight variations in traits give some individuals a slight leg up over others in the struggle for existence. When these variations are heritable, Darwinian evolution ensues as gradual and incremental shifts in those traits over many generations. When inheritance is Mendelian, as Morgan had demonstrated, variation occurs in discrete jumps—peas have either yellow flowers or white flowers, but they do not shift gradually from generation to generation from white to off-white, to ivory, to cream, to corn silk, and finally to yellow. Traits also arise sporadically and whimsically, not just as sports in gardeners' plots, but in wild plants as well. Of particular interest at that time was the seeming proliferation of new "species" of evening primrose, which could arise in a single generation and sporadically, with no seeming correlation to environment, let alone as the result of selection. Mendelian evolutionary change therefore would have to proceed in fits and starts: it would be "mutational," in a word. More importantly, it would be mutation, not natural selection, that drove evolution. This contradiction was the sharp end of Morgan's critique of Darwinism, and it was the basis of a new school of non-Darwinian evolution he called *mutationism.*[*]

Morgan's mutationist theory of evolution was the foundation upon which our modern theory of evolution was built, what has come to be called the synthetic theory of evolution, also known as Neo-Darwinism. The emergence of Neo-Darwinism simultaneously represents one of the greatest triumphs of twentieth-century science and one of the greatest ironies. We will get to the triumph momentarily, but first let us dwell briefly on the irony. Mutationism was, at root, an *anti*-Darwinian theory of evolution. Morgan himself thought it to be utterly incompatible with any model of gradual and adaptive evolution, which in his mind included both Darwinism and Lamarckism. Yet what grew from the mutationist foundation that Morgan had so carefully laid was a theory of evolution that called itself Darwinian but in reality had little that was Darwinian about it. How Morgan's fundamentally anti-Darwinian strain of evolutionary thinking came to drape itself with the mantle of Darwinism was, in short, the equivalent of a scientific Great Switcheroo.[†]

---

[*] Others called it Morganism.

[†] Roald Dahl's 1974 risqué short story about a wife-swapping scheme gone awry was published originally in *Playboy* magazine. It reflected Dahl's jaundiced view of the perversity of life, which in the story was the unintended sudden switch or swap that leaves the mark bewildered as to what just happened. Dahl's real brilliance as a writer was how he managed to thread this jaundiced view so seamlessly into children's literature.

Now, I want to be careful about my meaning here. Roald Dahl's story "The Great Switcheroo" was a reflection on credulity, but it was also a tale of the power of internally consistent thinking to lead someone happily right into a trap: it is a tale of epistemic closure. My argument here is that much of the history of evolutionism in the twentieth century has the marks of a Great Switcheroo unfolding. Morgan inadvertently set the trap, but we have been leading ourselves deeper into it ever since, confident in our path but unaware of the trap that waits to be sprung.

The self-deluding aspect of a Great Switcheroo is the reason why we are able so casually to accept the flagrant tautology that sits at the heart of our modern thinking about evolution: that adaptation arises by selection of genes for apt function and structure (Chapter 1). We have no trouble accepting it, despite having been told—from kindergarten, practically!—that tautology is the most vacuous of logical arguments. Yet there it sits, at the heart of the Neo-Darwinian conception of evolution. Does this mean that the Neo-Darwinian synthesis is cheapened by tautology? Far from it, for the Neo-Darwinian synthesis stands as one of history's greatest intellectual achievements, easily on a par with the quantum mechanics revolution in physics that played out at roughly the same time.

By now, your head may be spinning with the seeming contradiction of what I've just said. Is Neo-Darwinism a self-delusion, or is it a brilliant resurrection of an idea—Darwinism—that had largely been written off? There is really no contradiction, though, as long as we look upon Neo-Darwinism as first and foremost a *mathematical* triumph. Now, for mathematicians, tautology is not the same problem as it is for the rest of us, because all mathematics is, to a degree, tautological: what else is an equation, after all, but a restatement of a premise?* We do not conclude from this, however, that all mathematics is invalid, because a mathematical tautology is not a cheap trick—it is an affirmation of the virtue of internal logic and consistency, even if (and this is the really important part) that logic leads to strange conclusions. What makes the Neo-Darwinian synthesis a great intellectual achievement is the surprisingly strange and beautiful development of its central tautology. What makes the Neo-Darwinian synthesis questionable, on the other hand, is the way mathematics drew evolutionary thought deeper into the Great Switcheroo.

Morgan left evolutionary thought with a deep problem, because mutationism contradicts Darwinism's central claim. The gene is an agent of stasis, not of change, and this means the gene cannot be an agent of Darwinian evolution. Evolution can proceed only if the agent of stasis, the gene, changes; and in the mutationist catechism, this occurs in leaps, not via small changes. A mutation is just suddenly *there*, absent in one generation and present in the next. Thus, it is not organisms per se that evolve; only the genes that specify them can evolve.

From the point at which this logic is accepted, a profound shift in viewpoint is required if you are to think critically about evolution. Evolution now becomes the story of genes in populations of the carriers of them. I may carry a gene within me, but so too may all the other members of the population of human beings with which I might possibly mate. Because I, as an organism, am ultimately irrelevant to the fate of that gene (except as I may pass it on to another

---

* For the first part of the twentieth century, there was a search for a complete theory of mathematics, led largely by the German mathematician David Hilbert. The rationale for this was to find a truly universal mathematics that could unite all branches of mathematics under a universal set of axioms. In the early 1930s, Kurt Gödel ended this notion with his incompleteness theorems, which showed that all branches of mathematics are based upon distinctive sets of axioms. Because any mathematical conclusion is derived ultimately from a set of these essentially arbitrary axioms, all mathematics is therefore, to some degree, tautological.

irrelevant carrier of it), the object of our evolutionary attention turns to the *assembly* of genes held by all members of the population, or the *gene pool,* to use the phrase coined by Darwin's wayward half-cousin, Francis Galton. This means that the genetics of evolution becomes a problem in the statistics of genes in populations. And this means that the battleground for Darwinism shifted to the statistics of populations of genes. For much of the early twentieth century, the battleground was not favorable to Darwinism.

This is illustrated by the early debate over the phenomenon of gene dominance and gene recessiveness. Recessive genes, in the Mendelian sense, are genes whose expression is masked by the presence of a so-called dominant *allele* of the gene. A simple example of such a dominant-recessive gene is given by cystic fibrosis (CF), a disease of the lungs and kidneys. CF is caused by a mutation of a single gene, the CF gene (although there are at least seven different mutations of the CF gene that can cause the disease). The CF gene specifies a protein that sits in membranes and controls the transport of chloride ions across them. When the CF gene is defective, the protein it specifies no longer functions, with many of the devastating symptoms of the disease to follow—the excessive thick mucus in the lungs, the formations of cysts in the kidneys and lungs, and so forth. However, each person carries *two* copies of the CF gene, one from each parent. If one of these is defective and the other is functional, the disease may not occur. In this sense, the mutated gene, though present, is masked by the functional gene: the mutated gene is said to be *recessive,* while the normal gene, which encodes a fully functional protein, is said to be *dominant.* This means that the disease will occur only if there is no normal gene present to "dominate" the recessive allele, that is, if there is no gene to specify a functional chloride transport protein. In other words, if there is at least one copy of the normal CF gene so that chloride transport works normally, the disease will not be present, even if the person "carries" one copy of the defective CF gene. This can occur at odds of three-in-four, if both parents are carriers of the mutated CF gene. If only one parent carries the defective gene, no offspring will have the disease, although there is a one-in-four chance that a child will carry the defective gene.

We now know pretty well the nature of that relationship between dominance and recessiveness, but that was not the case in the early years of the rise of Mendelism. The Darwinian explanation for recessiveness was to propose a kind of natural selection among genes rather than individuals. Perhaps, the Darwinists argued, dominant and recessive alleles were not indifferent to one another's presence in a cell, as the Mendelians thought, but were somehow locked in a competitive struggle, so that the dominance of the dominant allele was the outcome of a selection struggle among the alleles. If so, then one would expect to see dominant alleles driving the recessive alleles eventually to extinction. This theory was known as *genophagy* ("gene-eating").*

Genophagy as a theory was fine and clever, but it was not something that could easily be tested experimentally. Here, statistical theory came riding to the rescue, in the person of G. H. (Godfrey Harold) Hardy, Sadleirian Professor of Mathematics, Trinity College, Cambridge.† Though Hardy was not a biologist, he enjoyed playing cricket with one, Reginald Punnett, who was an early pioneer in experimental genetics, and as it happened, was immersed deeply in the genophagy controversy.‡ Over their games on the Trinity College Backs, Hardy and Punnett would chat with one other about their respective interests. During one game, in 1908, when Punnett described the gene dominance problem to Hardy, Hardy

---

* Note also the complete absence of the organism from this whole argument. It was genes, not organisms, that were locked in Darwin's "struggle for existence."

† Hardy is profiled vividly in *The Indian Clerk* by David Leavitt (Bloomsbury, 2008).

‡ Undergraduate biology students will be familiar with his "Punnett square," a tabular device that could be used to predict gene frequencies that would result from crosses. I used a Punnett square to sort out the types of gametes one would expect for the two hypothetical traits of "stripy" and "color."

saw immediately that the whole genophagy idea had to be nonsense. He saw this not because he had any credible opinion on the biology of the matter, but because he clearly saw the problem as one of elementary permutations and combinations. Hardy then dashed off an equation showing, as a general case, that the frequencies of dominant and recessive alleles of a gene would not change from generation to generation, a constancy that came to be known as the Hardy-Weinberg equilibrium.* Genophagy was mathematically impossible, and therefore had to be biologically impossible.

As it turns out, Hardy was wrong on this. Since Hardy and Punnett, we have learned quite a bit about the evolution of genomes, and it turns out that quite a lot of competition is going on between genes.[1] Snippets of DNA wander from chromosome to chromosome, suppressing their new neighbors or entering into unholy alliances to win over other gene coalitions. Entire genomes churn and bubble, splitting off into new chromosomes or merging into larger chromosomal coalitions. Sometimes entire chromosomes and all the genes riding on them are being forced to extinction, as seems to be the case with the chronically puny Y chromosomes of male mammals.[2] Competition between genes can also manifest at higher levels of organization: differentiation of cell types, even tumors, are thought by some to be the means whereby genes can find expression independent of competing genes in other cells.[3] Life itself, it seems, cannot be easily corralled into a beautiful mathematical tautology: mathematical impossibility is not the same as biological impossibility.

In Hardy's time, in the first decade of the twentieth century, biologists did not have the benefit of our wonderful hindsight, of course. While Hardy's equation effectively beat back the genophagy

---

* For some years, this equation was known as Hardy's Law, until it was discovered that the German mathematician Wilhelm Weinberg had derived it independently several years previously. Consequently, the equation is properly known as the Hardy-Weinberg equation, and the constancy of gene frequencies it depicts is known as the Hardy-Weinberg equilibrium.

challenge to mutationism—there was no natural selection among genes—it did not provide much in the way of a constructive program for Darwinian evolution: it was an affirmation of the mutationist assertion of genes as agents of stasis, not change. Not only had Mendel shown that the nuggets of heredity jumped unchanged from generation to generation among individuals, but Hardy had also shown that their representation in populations would continue unchanged from generation to generation as well.

———•———

The problem for Darwinists at the time was how to rescue Darwinian evolution from the twin tyrannies of Mendelian mutationism and Hardy's mathematical tautology. This seemed to be an insurmountable problem, and there the situation sat for roughly a decade until three remarkable men came along and solved the problem: Ronald Aylmer Fisher (1890–1962), Sewall Green Wright (1889–1988), and J. B. S. Haldane (1892–1964). Together, independently, and all roughly simultaneously, they saved the day for Darwinism, and they did so by turning the tables on the Hardy-Weinberg equilibrium. Their achievement came to be known as the *genetical theory of natural selection,* also known as the Neo-Darwinist synthesis.

The three men came to the problem from very different backgrounds. Haldane was the aristocratic polymath—iconoclastic, Oxbridge, London, and trenchantly Marxist. Fisher was middle class, conservative, High Church, and spent his most productive years at an agricultural research station in Hertfordshire before returning to Cambridge later in his career. Wright, meanwhile, was a petit bourgeois Midwesterner from the United States.

Of the three, the most significant mathematical advances came from Fisher and Wright. (We will hear more from Haldane in a subsequent chapter.) Both Fisher and Wright were brilliant mathematicians, and both had come to natural history, biology, and evolution via very different pathways. Ronald Fisher's father was a London-born fine arts dealer, and his mother was the daughter of a prominent London solicitor; she died when Ronald was only fourteen. In school, he proved to be a mathematical prodigy, which earned him a string of distinguished scholarships that supported him through his prep school and university education (a good thing since his father's art business had collapsed to shambles by the time Fisher could have used his father's income). After his graduation from Gonville and Caius College, Cambridge, Fisher found himself at loose ends, so he took himself off to Canada to work on a farm. This experience planted in him a lifelong interest in biology and agriculture, and he returned to England keen on starting a farm himself. That scheme fell through, so he worked for a time as a statistician with an insurance company, until the Bosnian gangster Gavrilo Princip pulled the trigger on Archduke Franz Ferdinand and plunged the world into the devastating Great War. Fisher eagerly sought to enlist in the army but was rejected because of his poor eyesight. He taught mathematics and physics for a time, until greater opportunities for academic work began to come his way. This landed him ultimately at the Rothamsted Experimental Station in Hertfordshire, where he found ample room to indulge his growing interest in statistics and eugenics. While there, Fisher did pioneering work in statistical analysis and experimental design; his reputation as a founder of modern statistics is as substantial as his pioneering work in evolution. In 1925, he set down this work in *Statistical Methods for Research Workers,* a handbook that still finds use. Five years later, he released his evolutionary magnum opus, *The Genetical Theory of Natural Selection.*

Sewall Wright was also a mathematical prodigy, born in Massachusetts to a striving academic father and a sweet-tempered and liberally educated mother. Wright's father was a renowned political scientist and writer, but as a parent he was apparently somewhat distant from

Sewall, favoring Wright's more like-minded younger brother, Quincy (who went on to become a distinguished political scientist).* Sewall was a diffident and quiet young man who was disposed, like Ferdinand the Bull, to dreamy wanderings in fields.† This tendency forged a natural bond with his mother, who was a keen amateur naturalist and who encouraged her son in his pursuits of natural history. While Wright was still a boy, his ambitious father fell out with the trustees of the Massachusetts college where he taught, so he quit his job and took his family off to Illinois to take up a new position at Lombard College, a Universalist college in Galesburg.‡ When Sewall was ready for college, Lombard was the obvious place for him, so he matriculated there, planning to study mathematics. Instead, he found his intellectual feet under the tutelage of Wilhelmine Entemann Key, one of the few women in the United States at that time to have earned a doctorate in biology. Key was an avid naturalist, and Wright found in her someone who affirmed the attraction to natural history that his mother had nurtured.§ After graduating from Lombard, Wright went off to graduate school at Harvard to earn his own doctoral degree, studying inheritance of coat colors in mammals. After a short stint as a researcher at the U.S. Department of Agriculture, Wright took up a position at the University of Chicago, where he remained for nearly his entire career. Wright was a prolific researcher. Unlike Fisher, who made an early splash through his magnum opus *Genetical Theory,* Wright laid out his ideas in his four-volume *Evolution and the Genetics of Populations,* which he brought out over many years.

Fisher and Wright's achievement was two-pronged. One prong

---

* Wright's third brother, Theodore Paul, grew up to be a distinguished aeronautical engineer.

† Ferdinand is the title character in Munro Leaf's 1936 *The Story of Ferdinand,* illustrated by Robert Lawson (Viking Press) and popularized by Walt Disney. Ferdinand preferred to spend his days in the field smelling flowers rather than butting heads with other bulls, as they prepared for the bull ring.

‡ Among Lombard's distinguished alumni was the poet Carl Sandburg.

§ Wright maintained a warm correspondence with her throughout her life.

was dazzlingly successful: reconciling what seemed to be the irreconcilable tenets of Mendelian versus Darwinian evolution. The second was less successful: bringing Darwinian adaptation back into the mix after it had been unceremoniously shoved off the stage by the mutationists.

Turning first to the successful bit, we see that Fisher and Wright essentially rewrote the equation for the Hardy-Weinberg equilibrium to account for variations of fitness, which they defined as the likelihood of reproduction. We have been talking a lot about coat color, so let's frame the problem in those terms (Figure 7.3).* One can spin all sorts of interesting scenarios where coat color can affect fitness. Perhaps a light-colored coat would be useful in, say, an open habitat, whereas a dark coat color may be beneficial in a shadier, darker environment. In both instances, the matching of coat color with environment might provide camouflage for prey or added stealth to a predator. In contrast, dark coat colors in open habitats or light coat colors in shady habitats might be a disadvantage.

Imagine now a scenario where the environment gradually changes from open and lighter conditions to darker and shadier conditions (the vertical axes in Figure 7.3). This could come about in any number of ways. A population of light-colored animals could expand its range, migrating for some reason from open grasslands to forests. Populations of light-colored animals could find themselves at the fringes of their usual grassland habitats, allowing them to more frequently venture in dark and shady forested habitats. Or forested habitats could begin to encroach upon grasslands, effectively the new habitat coming to the animals. It scarcely matters which environmental scenario we're talking about. What is important is that there is a gradually developing mismatch of the prevailing coat color with the environment in which that trait occurs.

---

* The genetics and heritability of coat color was, in fact, a lifelong interest of both Fisher and Wright.

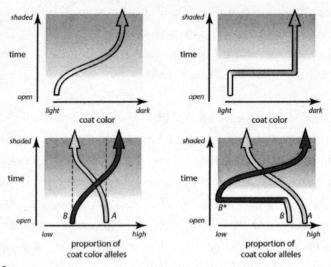

Figure 7.3

Scenarios for evolution and adaptation as regards coat color. *(upper left)* In Darwinian evolution, a continuously variable trait shifts gradually over time in response to a changing environment. *(upper right)* In Mendelian evolution, there is an abrupt mutation in a gene that may or may not result in an apt fit between organism and environment. *(lower left)* Fisher's and Wright's model supposes that adaptation results from a change in frequency of particular alleles in a population of genes. *(lower right)* Mutationism is reconciled with Darwinian evolution through modification of frequencies of newly mutated alleles.

One would expect that this change of environment would entail evolution of a darker coat color over time. In a Darwinian scenario (see Figure 7.3, upper left), coat color varies naturally within any generation, so an ancestral population of light-colored animals will contain within it individuals that are darker-colored than the norm. If the darker-colored individuals compete better in the gradually darkening environment, and reproduce better, the result will be a gradual change in coat color over time as these darker-colored individuals are selected for reproduction and come to dominate the population. In this instance, it is adaptation—the aptness of the fit between environment and organism—that drives the evolution of the population.

In a mutationist scenario (see Figure 7.3, lower right), coat color darkens when a mutation occurs to darken the coat. There is no possibility of gradual adaptation in this case: dark coat color just appears when a mutation for dark coat color appears, with a sharp discontinuity between light coats and dark coats. In the mutationist scenario, chance determines the survival of the new dark-colored trait. Adaptation might come along after the fact, but adaptation does not drive the appearance of dark coats, mutation does.

Around the turn of the twentieth century, these two mutually inconsistent scenarios represented the whole game for evolution. There could be Darwinian adaptation, but with no credible theory of heredity backing it up, or there could be a credible theory of heredity that had no room for Darwinian adaptation.

Fisher and Wright broke this logjam by casting the problem in statistical terms (see lower left of Figure 7.3). Now, genes for dark and light coats existed in the aggregate gene pool of an interbreeding population of gene carriers. It no longer mattered what genes an individual carried; instead, it was the proportional representation of each gene in the population that mattered. If light coat color was prevalent, this signified the statistical prevalence of the light-coat-color genes compared with the dark-coat-color genes. If the environment darkened, this would change the odds of reproduction, and hence the odds that a gene could be transmitted into the future. The result would be a gradual increase in the proportion of the dark-coat-color genes in the population over time. Fisher and Wright's new calculus of evolution could therefore reconcile mutationism with Darwinism, something Morgan, and Hardy, and Weinberg, among others, all had thought impossible.

With this achievement, Darwinism no longer needed vivid metaphors like Alfred Tennyson's "Nature, red in tooth and claw" to describe it, or Lamarckian meanderings about soft versus hard inheritance.* Fitness could now be dispassionately and precisely expressed as the tendency of an allele to replicate. And adaptation was now the mathematical—the *tautological*—outcome of the differential replication of different alleles.

Once Fisher and Wright had pulled off their brilliant reconciliation of Mendelism, mutationism, and Darwinism—the Neo-Darwinist synthesis—there remained that nagging problem of adaptation. Where was *its* place in the Neo-Darwinist synthesis? Adaptation *was* there, but in a radically different—a tautological and therefore meaningless—form. Adaptation was now a statistical adjustment of frequencies of genetic variation in aptitude. Some variants might code for apt function, while others might code for less aptitude. Adaptation was now no longer a striving for aptitude as it long had been thought to be, but a lottery: you either had it or you didn't, and if you didn't have it, your evolutionary fate would not have a happy ending.

We could at this point just throw up our hands and say that's that—it's intellectually sound and we might as well pull up our socks and just accept it. This is the argument of the Four Horsemen of the Evocalypse: accept that your life, or any life for that matter, has no meaning and no purpose, and just suck it up. It is to the credit of both Fisher and especially Wright that they did not take this approach. They would spend much of the rest of their lives trying somehow to

---

* The phrase "nature red in tooth and claw" comes from Tennyson's poem *In Memorian A.H.H.*, which explored the conflict between the Christian ideal of nature organized by God's love and the seeming brutal reality of nature itself. The phrase appears in Canto 56:

"Who trusted God was love indeed
And love Creation's final law
Tho' Nature, red in tooth and claw
With ravine, shriek'd against his creed."

Figure 7.4
The Blue Fairy. Adapted from *La Fata dai Capelli Turchini*
(The Fairy with the Turquoise Hair), as illustrated by C. Chio-
stri for Carlo Collodi's *The Adventures of Pinocchio*.

work adaptation meaningfully back into the bleak Neo-Darwinism
they had created. They were largely unsuccessful in this, and the
failure deservers a chapter of its own, to come. But before I close this
chapter, I want to make just one point. It involves Pinocchio.

We are all familiar with Carlo Collodi's famous tale of the
puppet boy who wanted to become a "real" boy—at least we are
familiar with the bowdlerized and Disneyfied version of the tale.*
Collodi's tale itself is quite dark, with Pinocchio left decapitated and

---

* *Pinocchio* was released in 1940, after the 1937 success of Walt Disney Production's first animated
feature film, *Snow White and the Seven Dwarfs*. Carlo Collodi's original fairy tale was *The Adventures of
Pinocchio* (*Le avventure di Pinocchio*), published in 1883.

hanging in a tree for his willful misbehavior and defiance. Only after Pinocchio has reached this unhappy end does the Blue Fairy* come to the ghost of Pinocchio in a dream to lead him through the reasons for his fate. But Pinocchio's ghost is blind to his misdeeds, and when he tries to lie his way out of the reality with which the Blue Fairy is confronting him, his wooden nose grows longer. Still, the Blue Fairy is merciful, bringing in woodpeckers to whittle down his nose in the hope that Pinocchio will see his folly. Eventually he does, and only then does the Blue Fairy turn Pinocchio into the "real" boy he had longed to be.

Collodi's story is a conventional one of taking responsibility for one's life, but there is another side to it—that of self-induced blindness to reality. Pinocchio could not see how his behavior had led him on the path he followed. His behavior seemed to him always to be the right thing to do. What else is a puppet boy to do but play tricks on people? Wasn't that what all puppets did? Pinocchio's world was logical and internally consistent. In other words, Pinocchio's world was epistemically closed.

Modern Darwinism has worked itself into a similarly closed universe, where all seems right and logical. Yet modern Darwinism is not where it is because its tenets have been objectively proved; the history of evolutionary genetics is a tangled one, with many competing narratives that reconcile adaptation and heredity. Modern Darwinism—Neo-Darwinism—has become the supreme narrative because it largely ignored (and sometimes suppressed) these alternative and competing narratives.[4] The price of winning this competition has been to pull off a reverse Pinocchio. The biology of the nineteenth century saw the meaning of being a "real" boy, that is, a boy filled with vitality and striving, but the physics envy that drove the Neo-Darwinian revolution drained this vitality away. In short,

---

* Disney's Blue Fairy was adapted from Collodi's Fairy with Turquoise Hair.

biology turned to the Blue Fairy and entreated her to make him a wooden boy. She has largely complied, leaving life to be poked and prodded by e. e. cummings's "naughty thumb of science."

This is the unhappy fate of modern Darwinism, but there is a cause and it is easy to see. It was the elevation of heredity as the sole driver of evolution, which, as we have seen, was an anti-Darwinian idea. But for Fisher and Wright (and Haldane), we would not even call this theory Darwinian. Yet try as they did to work adaptation back in, none of the principal architects of Neo-Darwinism—Fisher, Wright, or as we shall see, Haldane—could quite find the mathematics for doing so, because evolution had, by their hands, been turned into a wooden boy.

There is no malice or stupidity in any of this, but it is useful to know what has continued to draw modern Darwinism into its Great Switcheroo, that is, the confident wandering into the trap of a self-induced divorce from reality. Since heredity became the sole beacon guiding modern Darwinism, what of heredity itself? Perhaps there is something amiss about our conception of it. Heredity is a form of memory, to be sure, but just what do we mean by this? The Neo-Darwinists, being materialists, treated hereditary memory as objects, gene "things" that specified function. Perhaps this is where evolutionism in the twentieth century went wrong.

Let me pose the question this way: given that life is an ephemeral dynamic disequilibrium, as discussed in Chapter 2, perhaps hereditary memory is too? Rather than memory being replicable object-code, as we assume the gene to be, perhaps memory is more process than object, more disequilibrium than stasis, more verb than noun. This leads us to a radical idea, which we will now begin to develop— that evolution is driven not by natural selection, but by *homeostasis,* and the implied striving and desire that homeostasis implies.

8

A MULTIPLICITY OF MEMORY

Memory is never a thing alone; it is always something else, a "strange Bell—Jubilee, and Knell," to quote Emily Dickinson. Memory is dynamic, fleeting, always arising in association with something else: a smell, a sight, a rustling of leaves, a photograph that draws out remembered moments from long ago. For that matter, memory is never really a *thing* at all. We may strew tokens of our memories all around us—trinkets, letters, books—that evoke memories in ourselves and in others, but these tokens are not the memories themselves. Tokens are things we can hold in our hands; memories are living experiences.

In the Neo-Darwinist conception, evolution is supposed to be governed by a kind of memory, knells that peal from the past to this day, embodied, we are told, in a molecule—deoxyribonucleic acid, or DNA—that encodes the memory in a sequence of nucleotides. The fact that we understand the encoding of hereditary memory in DNA is the gift of the molecular revolution that consumed the science of biology through the latter half of the twentieth century, culminating in a fully material definition of the gene. That gift illuminated something profound about evolution, for it showed how codes that once rang harmoniously and sweet could be remembered, while those that sounded with a dull clunk could be forgotten. This is the modern triumphant conception of natural selection, for in those DNA codes are August Weismann's ids incarnate, the genes of Gregor Mendel and of Thomas Hunt Morgan unmasked, the equations of Ronald Fisher and Sewall Wright written in chemistry.

At the risk of sounding a dull clunk myself, allow me to pose a question. Is it the codes that have really guided the evolution of life? How we answer this question turns on what we think the codes are. Are they the hereditary memories themselves, or are they the tokens of hereditary memory? The distinction is important.

———— • ————

To help clarify the distinction, I would like to introduce you to my mother, through a memory token of my own, that is, a photograph taken of her shortly before her marriage to my father (Figure 8.1). I introduce her because I want to use my memory of her as the starting point for an exploration of what memory means. This is very important, because evolutionary memory—heredity—has, ever since Morgan, been placed at the center of how we think about evolution. Indeed, our thinking is even more circumscribed than that, because the logic of genetic natural selection points to there being only one form of hereditary memory that matters—the nucleotide sequence code of the gene. It is important, therefore, to be very clear about what memory is.

When I look at the photograph of my mother from long ago, the memory of her begins with an infinitude of photons streaming into my eye from the photograph—either reflected light from a paper or emitted light from a screen of some sort. My cornea and lens focus these photon streams into a facsimile image onto the retina lining the back of my eyeball—two slightly different images actually, because I

Figure 8.1
Lucille Vawter Busby Turner, my mother, as a
young woman.

have two eyes, and each looks at the photograph from a slightly different angle. My retina is a sheet of innumerable light-sensitive cells, called *photoreceptors,* which respond to illumination with a slight change in the tiny voltages that exist across their membranes. The pattern of electrical tingles shaped by the projected image is then conveyed to other cells within my retina, which talk among themselves about the messages coming to them from the photoreceptors, and they respond with little tingles of their own. The pattern of these tingles represents a facsimile of the image projected onto the retina, which is then conveyed to my brain via a communications trunk line (my optic nerve). Once in the brain, the facsimile is processed through several complicated steps until it ends up tickling a community of nerve cells inhabiting the back portion of my cerebrum (the occipital cortex, as it is called).*

Complicated as the conversation is so far, it is really only beginning. The cells in my occipital cortex then begin to talk to all of

---

* For interested readers, I have outlined the visual cognition system in more detail in *The Tinkerer's Accomplice.*

the other cells in my brain, and from those conversations come the flood of memories that rise into my consciousness—some dim and indirect, such as memories of her own mother (my grandmother, Mamaw) whom I remember only from photographs, stories, and one short time in her presence when I was a very young child. Some memories are more vivid, such as of my maternal aunts and uncle, with whom we would stay on our yearly visits to my mother's east Texas home. One memory in particular stands out, of working one summer on my uncle's tomato farm, of the intense heat of the east Texas summer, and of the sumptuous midday meals washed down with gushers of the sweetened iced tea that is the trademark beverage of the American South. Some memories are whimsical, such as of our drives from California to Texas, before cars had air conditioners, with me wearing wet washcloths on my head as we drove through the Mojave Desert. Some memories are proud, such as remembering my mother as my father's vibrant business partner, keeping the books and simultaneously soothing and fighting off bankers and creditors, building with him a business that eventually reached a million dollars in annual revenue—quite a lot of money in the mid-1960s. Some are painful, such as her later struggles with alcoholism and divorce, which culminated in the destruction of our family and everything that my parents had built up over their twenty-five years of marriage and partnership. Some are hopeful, such as her brave attempts at rehabilitation, recovery, and redemption, which she eventually won. And some are tragic, such as her early death from breast cancer at the age of fifty-six, my own anger at her, and my guilt and grief at my own shortcomings as her son. You will have your own memories of your own mother, of course, and I could share many more of mine, but the simple point I wish to make is this. How trivial is the token of her memory, a pattern of ink splashes laid out on a square of paper! How rich is the flood of memory the token evokes! And how vast and unknown is the gulf between them!

I tell this story because it underscores, for me, two fundamental problems with our current conception of hereditary memory. The first is that in our quest to understand the material foundation of the gene, we have conflated the *token* of hereditary memory—a snippet of DNA—with hereditary memory itself. Given the vast gulf that likely exists between them, it is difficult to see how we can possibly understand memory, in whatever of many forms it might take—neural, genetic, evolutionary—from memory tokens alone. Yet that is precisely what the gene-centered legacy of Morgan, Fisher, and Wright asks us to do.

The second shortcoming is that the gene-centered conception of hereditary memory truncates memory's full dimensions. We regard memory as an evoker of the past. The memory token of my mother certainly does that, and it is certainly the case with the memory tokens of naturally selected genes. Those snippets of DNA exist today, after all, because they somehow helped to shape apt functions in the past. But this backward-looking perspective, valuable as it is, misses memory's other essential dimension: it is not only a token of the past, but a harbinger of the future. We have a natural tendency to look backward to the time a memory was formed, a remembrance of things past, to use that overworked phrase. Yet at the time any memory is formed, it biases the future to ensure that it unfolds in a particular way out of many possible futures, including ensuring there is a future me who will look backwards in time. The memory of my mother formed in the past sets up the future so that I will react in the way I do, many years after. This future-looking perspective does not sit comfortably in modern biology, because to look forward is to admit the possibility of foresight, purposefulness, and desire shaping evolution. By keeping our gaze fixed firmly backward, we can safely keep our minds turned away from that difficult perspective. Can we afford to continue to do so? I think not.

That gulf between memory token and memory itself beckons us now to wade in. We will be following a path that at times will become tortuous and hazy, but I will ask you to stay with me, because the path is marked all the way by threads, metaphorical in places, literal in others, clear in places and tangled elsewhere, but always there for us to follow if we look diligently enough. So, allow me to be the Virgil to your Dante, for there is a surprising conclusion about memory awaiting us at the threads' terminus.

Our first tentative step, really just a wetting of our ankles, starts with those photoreceptors in the retina. There are two types of photoreceptors in our eyes, but both make the point equally, so I'll focus on just one, the *rods* (Figure 8.2).[*][1] These make up the bulk of the photoreceptors in our retinas, and they are a marvel, so extraordinarily sensitive to light that they can register the interception of a single particle of light.

The rods act as photoreceptors because they snag photons streaming into our eyes and convert their energy into signals that our brains can understand. This ability rests in the rod's "light antenna," the so-called outer segment. This is a long membrane-bound cylinder, packed with tiny flattened sacs of internalized cell membrane, a few dozen of them stacked upon one another like coins. Sitting at the base of the outer segment is the first anchor to the thread that will lead us deeper into the gulf: a bizarre little structure called the *basal body*, a small bundle of nine short microtubules bound together like the sticks of the Etruscan axe called a

---

[*] The other type of photoreceptor in the vertebrate retina is the *cone*, supposedly responsible for color vision, but what I have to say applies equally to cones and rods. A variety of other photoreceptor types are to be found among other animals or cells, the most notable being those of arthropods (insects, spiders and mites, crustaceans, etc.), which are derived from small fingerlike projections of the cell membrane called *microvilli*.

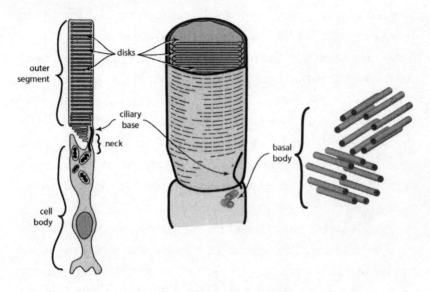

Figure 8.2

The rod photoreceptor. *(left)* Whole rod photoreceptor, showing outer segment with membrane disks within and cell body and neck with ciliary base extending into the outer segment. *(middle)* Cutaway view of outer segment, showing how disks are stacked like coins. The basal body is located at the base of the outer segment. *(right)* Diagram of basal body, showing basic architecture of two perpendicular bundles of nine microtubules each.

fasces. From the basal body, a tiny fragment of a cilium threads its way up into the outer segment.

The basal body, for all its weirdness, is actually a common feature in many of the body's other cells.[2] So it is not its presence in the rod cell that is odd, but rather the context of its presence. Usually, basal bodies are found anchoring the flagella and cilia (or as they are more properly named, the *undulipodia*)—whiplike fibers that beat back and forth to move fluids about—that decorate many cells. The airways in your lungs, for example, are lined with billions of undulipodium-bearing cells, busily sweeping mucus up the airways and out, and with it whatever dust, bacteria, and other trash the mucus has accumulated from the air pulsing through the pipes. Single cells, like sperm, can also sport undulipodia, but there the purpose is locomotion, moving cells through liquid, basically the reverse of what undulipodia in the body's fixed cells do.

So, here's a puzzle. The basal body at the base of the photoreceptor suggests that the outer segment is actually a modified undulipodium. Why, we must ask, is something that is usually associated with moving fluids about also found associated with a structure that senses light? The puzzle only deepens when we begin to look at other sensory cells within our bodies: whether it is the cells that let us smell, taste, hear, or feel, our sensation appears to begin, as it does in photoreceptors, with a modified undulipodium.

Several threads radiate from the intriguing anchor point of the basal body. Most radiate whence we've come, connecting the ability to move in the environment (or to move the environment) with the ability to sense the environment. That in itself is a pretty idea that excites admiration and wonder, and we will come back to it shortly. For now, there is another plot thread that leads us, not back to the photoreceptor, but deeper into the gulf. That is the thread we shall follow.

———•———

That thread leads us from the undulipodium to, of all places, a remarkably diverse kingdom of single-celled creatures, the *protists*.[*3] The protists inhabit water: ponds, oceans, drops in the clefts of orchids, tiny bridges of water suspended in the tangled threads of moss and molds, the vast sloughs of rivers, and the smaller rivers of liquid that course through organisms' bodies. The protists represent something more, though: they are the first *eukaryotes,*[†] a radically

---

* More properly called the Protoctista. Not all protists are solitary; many are colonial.

† The name means "true seed carrier," which refers to the prominent nucleus that sits, like a sun, in the center of the cell. The animals (including ourselves), plants, and fungi are descendants in various circuitous ways from the original eukaryotes, making those the major eukaryotic kingdoms.

new form of life that appeared on Earth about one and a half billion years ago.*[4] The thread that led us to them caught our eye because the undulipodium is the protists' evolutionary invention. Nothing like it had existed before the protists undulated their way on to the evolutionary stage. Once on stage, they changed the course of evolution radically.

From this new vantage point, we see strewn about our feet a myriad of new threads that beckon us now to follow. One prominent thread is the memory token of DNA itself, the long strands of nucleic acid that are the chromosomes. The thread of DNA connects the eukaryotes to the broadly defined bacteria, the *prokaryotes*.[†] The prokaryotes also have chromosomes, although these are quite different in form from the eukaryotic chromosome. In the eukaryotes, the chromosomes are long open-ended threads. In contrast, among the prokaryotes, the DNA thread is looped into a single-stranded ring-shaped chromosome called a *chromonema*. Eukaryotes' open-ended chromosomes are also decorated with a fabulous array of proteins and other supporting structures called the *histone complex*.[5] Prokaryotic DNA, in contrast, is decorated with spartan simplicity. Keep this decoration in mind: it will be important momentarily.

Prokaryotes and eukaryotes also contain their respective DNA threads within themselves in different ways. The eukaryotes envelop their chromosomes in a nucleus, containing them in a double-membrane sheath. Prokaryotes do not have a nucleus: their chromonema floats unsequestered in the cell. The nucleus is only the most prominent of a large set of differences with the eukaryotes, however. Eukaryotic cells are larger, by about an order of

---

* Generally, the origin of the eukaryotes is placed somewhere between 1.6 and 2.1 billion years ago. The older estimate is probably the safer, as there is fossil evidence of primitive eukaryotes existing at 1.6 billion years, which implies an origin earlier than that.

† The name means "before the seed carrier"; at least two bacterial kingdoms are included: the Eubacteria and the more primitive Archaea. The prokaryotes do not contain their DNA in a nucleus as the eukaryotes do.

magnitude. They also contain within them large organelles, called *plastids*, that are absent from the prokaryotes. One such plastid is the familiar mitochondrion, which harnesses oxygen to power the cell's metabolism; nearly all eukaryotes have mitochondria,* but prokaryotes lack them.[6] Another plastid is the chloroplast, found in green plants and photosynthetic protists, which captures energy in light to power the synthesis of sugars and other nutrients, farting out oxygen gas as a waste. The typical eukaryotic cell also contains within it a system of membrane-bound sacs called the *endoplasmic reticulum*. All of these—plastids, nucleus, endoplasmic reticulum, nuclear membrane—are bound up in a cobweb matrix of tiny tubules and filaments composing the *cytoskeleton*, the cell's internal scaffolding. The eukaryotic cytoskeleton is more than mere scaffolding, though. It is dynamic and ever-changing, pushing and prodding the eukaryotic cell into a variety of interesting shapes that the prokaryotes simply cannot physically manage.[7] The cytoskeleton also holds the plastids into place, manages the shape and function of the endoplasmic reticulum, and—significantly—controls those threadlike tokens of hard inheritance, the chromosomes.

The cytoskeleton's control of the chromosomes is dramatically evident during the division of the cell. Prokaryotes generally reproduce by simple fission: the chromonema is copied (replicated, to use the proper term), and the cell membrane grows and pinches off, making two new cells, each with its own copy of the original chromonema. In contrast, the cellular reproduction of eukaryotes, called *mitosis*, is a more complicated affair. It is an elaborate dance of chromosomes, all managed by a spectacular spray of microtubules, the

---

* I have to say "nearly all" eukaryotes because some primitive protists and some fungi either lack mitochondria altogether or employ an anaerobic version of the mitochondrion called the *hydrogenosome*. In the absence of oxygen, the hydrogenosome can employ protons (the hydrogen ions that make a solution acidic) to produce hydrogen gas. Hydrogenosomes can also function in the presence of oxygen, in which case the final waste product is hydrogen peroxide.

spindle apparatus, which radiates from two focal points positioned at opposite sides of the dividing cell. These microtubular sprays grab the chromosomes, push them this way and that, calling the dance until the chromosomes are finally sorted into two sets on opposite sides of the cell. Only then does the cell membrane start to fold in on itself, dividing the once-singular cell into two. If we look closely at the source of the microtubular spray, we see that it contains an organelle known as a *centriole*—which, upon close inspection, reveals itself as akin to a basal body!

Out of the tangle of threads, literal and metaphorical, we have been following and combing through, finally the plot connects: the spindle, the undulipodium, the cytoskeleton, mitosis, the photoreceptor, the very origin of the eukaryotes themselves—all seem to wrap together at this strange little barrel of microtubules.[8]

Radiating from the centriole, we now begin to see a tangled multitude of other interesting threads we might follow in the unfolding plot. That we are able even to see them we owe largely to that astonishing biologist Lynn Margulis (Figure 8.3). Margulis may be best known for serving, as Thomas Huxley did for Darwin, as the "bulldog" for James Lovelock and his Gaia theory.* She also is renowned for her provocative thoughts on the origin of the eukaryotic cell, and it is those, not Gaia, that interest us here.

Her scheme for the origin of the eukaryotes, which she called the *serial endosymbiosis theory* (SET), proposes that the eukaryotic cell emerged from a series of symbiotic associations between various types of prokaryotic ancestors.[9] Thus, the eukaryotic cell was

---

* Gaia theory is James Lovelock's controversial idea that the living Earth is essentially a living organism. In Lovelock's conception, the Earth does not simply harbor life, Earth is itself alive.

Image 8.3
Lynn Margulis (1938–2011).

not so much the product of a changing lineage of familial descent, but was the result of a series of mergers between diverse families of microbes. To put it another way, the origin of the eukaryotic cell was not so much Darwinian as it was Lamarckian.[10] As you might expect, it's a controversial idea.[11]

The first element of SET that Margulis brought into focus was the origin of the *plastid organelles:* the mitochondrion and chloroplast. In the scheme of SET, the plastid organelles originated as bacteria that took up residence within other prokaryotes,* driven there by metabolic necessity. The necessity arose because bacteria are notoriously inventive in their chemical metabolism. Among their more mischievous metabolic tricks was inventing photosynthesis, that is, the ability to harness the energy in sunlight to split apart water and to use the fragments to hydrate carbon dioxide and make carbohydrates—to make sugars, to put it simply. This little trick of metabolic legerdemain, hit upon about two billion years ago, tapped the bacteria that

---

* That is where the term "endosymbiosis" comes from—literally, symbiosis within. It is a common habit among bacteria.

had mastered the trick into a virtually infinite source of energy (sun-light) and "food" (carbon dioxide and water). This had far-ranging implications, among them precipitating the Greatest Ecological Catastrophe of All Time (GECAT). In a relatively short span of time, these photosynthetic bacteria flooded the world with the waste prod-uct of photosynthesis (oxygen gas), "rusting the Earth" and con-verting a mostly carbon dioxide–rich atmosphere into the strongly oxidizing mix of oxygen and nitrogen gas that we presently breathe.[12] The oxygenation of the Earth, seemingly so benign an outcome to ourselves, precipitated the GECAT because oxygen gas was a deadly poison for most of the bacteria that existed then. And those nouveau riche photosynthesizers were pumping it out in gushers.

Metabolic inventiveness may have initiated the GECAT, but the prokaryotes' metabolic opportunism was also the salvation from it, for as the catastrophe was unfolding, some bacteria learned how to recombine that waste oxygen with hydrogen and detoxify it to water, and to use the energy released to power their own metab-olism.*[13] According to SET, the mitochondrion had its origin when one of these oxygen-detoxifying strains took up residence within another bacterium that lacked this ability. It was, as they say, a win-win for all concerned, a kind of mutual scratching of microbial backs, if microbes can be said to have backs. The host, once imper-iled by oxygen, could now live in the newly emerging toxic world, thanks to the oxygen-detoxifying bacterium (the endosymbiont) liv-ing within. The oxygen-detoxifier, for its part, got a nice place to live and a ready source of "food" in the bargain. So compelling were the mutual benefits that the association eventually "took" in the form of the mitochondrion-containing cell.

---

* Saying this sounds rather glib, but bacteria seem endlessly inventive in finding novel fuels to power their metabolism, even to the point of being able to use rock as "food." To assert that bacteria simply invented a means of detoxifying oxygen is no less astonishing than asserting that bacteria simply invented a way to power their metabolism with light.

The origin of the plastid organelles was the first peculiar feature of the eukaryotes to be clarified by SET,* but this was low-hanging fruit that didn't really get anyone close to explaining the origins of the eukaryotes. The plastid organelles were the Johnny-come-latelies of serial endosymbiosis, probably the last in the series of associations that led to the eukaryotes, fully realized. The plastid organelles also represent only one of myriad distinguishing features of the eukaryotic cell, which include the nucleus, cytoskeleton, mitosis, and many others. Any credible theory of the eukaryotes' origin would have to account for them as well. The surprising account, arrived at after many years of painstaking work by Margulis and her colleagues,[†14] points to the nucleus, the undulipodium, and the cytoskeleton arising as three facets of the same early symbiotic association. Out of this symbiotic trinity arose what Margulis and colleagues called the last eukaryotic common ancestor (LECA), the proto-eukaryote that gave rise to all eukaryotes.

According to Margulis, that *Ur*-symbiosis was between a primitive bacterium that lived in hot acidic environments, similar to the *Thermoplasma* bacteria that inhabit hot springs, and a spirochete-like endosymbiont, both drawn together by a mutual interest in sulfur, of which more momentarily. The plot of the story unfolds as a nearly seamless weave of metabolism, locomotion, sensation, and memory, leaving at the denouement the undulipodium, the cytoskeleton, the nucleus, and the open-stranded DNA of the eukaryotes standing on the stage. There are many plot threads here: let's begin to tease them out.

Locomotion is the first thread we follow. Many bacteria use

---

* The conventional wisdom points to a so-called α-eubacterium as the ancestor of the mitochondrion and a cyanobacterium (sometimes improperly called a blue-green alga) as the ancestor of the chloroplast. This story has come under stringent criticism by Tim Cavalier-Smith. See Cavalier-Smith's "The Phagotrophic Origin of Eukaryotes and Phylogenetic Classification of Protozoa," *International Journal of Systematic and Evolutionary Microbiology* 52, no. 2 (2002): 297–354.

† Notably Ricardo Guerrero, Michael Doyle, Michael Chapman, and John Hall.

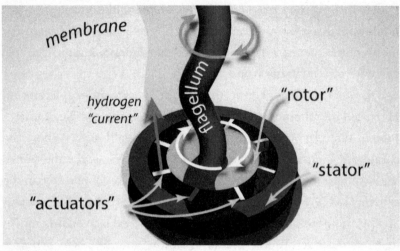

Figure 8.4
Schematic diagram of the bacterial flagellum.

flagella to move about their liquid environments. The word "fla-gella" (singular, flagellum) is unfortunate usage here because the flagella of bacteria and the flagella of eukaryotes are entirely differ-ent things.*[15] The bacterial "flagellum" is misnamed because it does not whip about, it actively rotates. It is a helical protein wire, a lit-eral thread anchored at its base to a rotor-armature-stator assembly embedded in the bacterial cell's membrane (Figure 8.4). These com-ponents of the bacterial flagellum are aptly named because they are reminiscent in structure and function of the same components in an electric motor.[†] A proton "current" (current is any flow of charge, remember, so positively charged protons carry a current when they move, just as electrons do) streams through a membrane-spanning "armature" that drives a rotation of the rotor within the membrane-fixed stator. As the rotor turns, the attached helical flagellum turns

---

* This is why "undulipodium" is the favored name for these locomotory devices among eukaryotes.

† The similarity is more than structural. The rotation is actually powered by an electrical current of sorts, not of the electrons that power our own electric motors, but of positively charged protons flowing across the cell's membrane.

with it, drilling through the liquid environment like a corkscrew and pulling the bacterium behind it. This is a very unusual design: the motor of the bacterial flagellum represents the only known example of a wheel in the living world.[16]

The eukaryotic flagellum, in contrast, is a cylindrical extension of the cell membrane that contains within it a long bundle of microtubules that are organized similarly to the basal body.[17] Where the basal body consists of a barrel of nine microtubules, the interior of the undulipodium contains nine paired microtubules plus two in the center. This "9 + 2" array, as the arrangement is called, is the undulipodium's motor. Locomotive power is provided by a sliding movement of the microtubule bundles past one another, which makes the motion of the undulipodium, well, undulatory, and quite a bit more versatile than that of the bacterial flagellum. The undulipodium whips back and forth, it oscillates sinuously, it twists itself into a traveling wave, but one thing it does not do is spin.

Even though the bacterial flagellum is limited in its motion compared to the eukaryote's undulipodium, bacteria have built a variety of interesting designs around this simple motion. At its simplest, the flagellum is a corkscrew, as just described. Sometimes, several flagella may intertwine to rotate synchronously, giving the bacterium a little more locomotive power and maneuverability. The design that interests us here is the flagellar design in spirochetes, those nasty, corkscrew-shaped bacteria that give us Lyme disease, syphilis, periodontal disease, and many other afflictions—and that also was one of the supposed partners in the LECA.[18] Spirochetes have two flagella, one anchored at each end of the corkscrew, contained within a space between the bacterium's cell membrane and an outer bacterial cell wall (Figure 8.5). Here is the really unusual design feature. When the flagellar motors are powered up, their action rotates the entire helical assembly—bacterium and cell wall—to auger its way through its environment: the entire cell, rather than just the

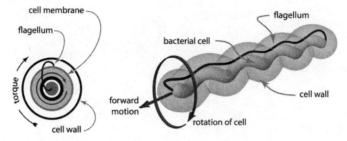

Figure 8.5

The spirochete locomotory system. *(right)* Perspective view of spirochete, showing the bacterial cell embedded into a cell wall. The flagellum is attached at both ends of the cell. Clockwise rotation of the entire assembly augers the cell forward, following the left-pointing arrow. *(left)* Head-on view of the spirochete, showing the bacterial cell (light grey circle) encapsulated in the cell wall. The flagellum is attached to the fixed rotor (dark gray center circle) and rotates the stator (the dark line outside of the rotor) clockwise, imparting a rotational torque to the entire cell.

flagellum, has become a large corkscrew. This allows the spirochete to drill through more viscous and denser media than the much narrower bacterial flagellum can by itself. This is one of the reasons why spirochetes are often pathogenic: they can drill more easily through the dense tissues and layers of mucus that organisms often deploy to hold microbial pathogens at bay.

So, that's the locomotion plot thread teased out. Let's now set that to the side for a moment so we can tease out the next thread in the story, that weird mutual common interest in sulfur. *Thermoplasma* breathe sulfur as we breathe oxygen (Figure 8.6). That is to say, *Thermoplasma* take elemental sulfur and combine it with hydrogen to make as waste the foul-smelling gas hydrogen sulfide (H_2S).* This produces energy, which the *Thermoplasma* harvest to power metabolism.

Now, there's an important thing to know about sulfur: it is chemically similar to oxygen and can do many of the same chemical things

---

* Hydrogen sulfide, the gas produced by rotting eggs, is produced by bacteria consuming the sulfur-rich albumin (egg white). It is also commonly produced in anaerobic wet sediments, which is why digging into mud flats can release a foul "rotten-egg" odor.

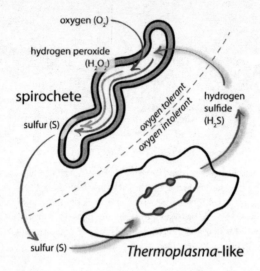

Figure 8.6

Cooperative sulfur metabolism of the *Thermoplasma*-spirochete association. The *Thermoplasma* releases hydrogen sulfide as a waste product. The spirochete consumes oxygen and produces hydrogen peroxide as an aerobic waste product. Hydrogen peroxide then combines with hydrogen sulfide to precipitate elemental sulfur within the spirochete cell wall.

that oxygen does.* Specifically, both sulfur and oxygen have a strong tendency to pull electrons toward themselves from other chemical bonds. Oxygen's pull is stronger than sulfur's, which is why oxygen reacts so readily with other molecules. But in a pinch, sulfur can do the trick as well, which is how *Thermoplasma* use sulfur. Just like oxygen can be combined with hydrogen to produce water,

---

* Certain classes of anaerobic bacteria use elemental sulfur as a metabolic analogue to oxygen. Both elements occupy the same column of the periodic table, and this means that both can serve as powerful oxidizers of metabolic substances. In eukaryotes and aerobic microbes, oxygen serves as the so-called final electron acceptor, drawing electrons through a complex chain of chemical reactions that produce ATP. In this instance, the oxygen ultimately reacts with hydrogen to produce water. Sulfur can also serve as a final electron acceptor, in which case the end product is hydrogen sulfide. It works the other way too. Photosynthesizers use energy in light to pull electrons away from water, leaving molecular oxygen as the waste product. This is what green plants do so effectively. Some bacteria pull off a similar trick, sometimes with light energy and sometimes with strong chemical potentials in high-acidity environments, to pull electrons from hydrogen sulfide, leaving elemental sulfur as the waste product.

$$O_2 + 2\,H_2 \rightarrow 2\,H_2O + \text{energy},$$

with energy released as a bang, so sulfur can play a similar role, producing hydrogen sulfide gas:

$$S + H_2 \rightarrow H_2S + \text{energy}.$$

The reaction details are less important for our story than what the reactions do for cells. By drawing charged electrons toward themselves, oxygen and sulfur both can power an electrical current, this one carried on electrons, which can be tapped to do work, in a similar way to any electrical appliance. When this happens in cells, the name we give to this work is *metabolism*. *Thermoplasma* therefore do with sulfur what our own cells do with oxygen: use it to power a flow of electrical current that does metabolic work.

There's just one rub. Once a sulfur atom has drawn an electron to it, it loses its power to draw more: it runs down, just as a battery does. To keep doing metabolic work—to keep living, to put it bluntly—the *Thermoplasma* have to get rid of the electron-sotted sulfide and bring in fresh electron-hungry sulfur. Just pooping the sulfide out is no solution, because that would fill up the surroundings with spent sulfide, fouling the nest, so to speak. So, not only does there have to be a continuous inward stream of sulfur to power the metabolic work, there has to be a continuous outward stream of sulfide away. When it comes to powering metabolism, the garbage truck is as vital as the grocery truck.

Here is where spirochetes enter the picture. Spirochetes use oxygen as *Thermoplasma* use sulfur—to draw in electrons to power metabolism. This is a very clever thing to do, but it means that spirochetes also have a waste problem, this one in the form of hydrogen peroxide waste that, if left alone, could do considerable damage to the cell. Just think of how hydrogen peroxide foams when we drip it onto an infected wound, each tiny bubble representing a bacterium being sent to its doom. A better way for spirochetes to deal with their hydrogen peroxide problem is to combine it with . . . hydrogen sulfide!—to produce water, hydrogen gas, and elemental sulfur. In spirochetes, the sulfur is deposited as a globule in the space between the cell and cell wall. Once the spirochete dies, this sulfur becomes part of the mud, ready to be consumed by sulfur-breathing bacteria—like *Thermoplasma*. The circle of life!

In microbial communities, this tiny circle of life plays out in layered muds called *microbial mats*.[19] Sulfide-consuming spirochetes will concentrate in a layer just slightly above a layer where the sulfide-producing *Thermoplasma* are abundant. Arranged in this way, the two microbial species can cycle sulfur between them, all powered by the downward trickle of oxygen from the atmosphere above. There is an added bonus. Oxygen is a deadly poison to sulfur-breathing bacteria like *Thermoplasma*. The layer of spirochetes above therefore acts as a protective filter to intercept oxygen before it can reach the *Thermoplasma* layer just below.

According to Margulis, the LECA supposedly arose in just such an environment, when a sulfur-breathing *Thermoplasma*-like ancestor (we will call it the TLA for short) closed this metabolic loop by bringing a sulfide-eating spirochete into its "body."[20] Now, the sulfur that normally sat sequestered within the spirochete's cell wall was brought directly into the TLA cell. The clever trick was keeping the spirochete's living function going inside the TLA, akin to how the duck in *Peter and the Wolf* kept quacking after the wolf had swallowed it whole.

What I have described so far is the conventional logic of endosymbiosis: bring another organism into your "body" and with it a new and complementary metabolism, as was the case for the mitochondrion. For the LECA, there was a wrinkle: the endosymbiosis was only partial. The spirochete's innards—its cell contents, including its chromonema—were drawn into the TLA cell, leaving the spirochete cell wall, the membrane, and flagellum assembly attached as a husk to the outside of the TLA cell membrane. Because the spirochete's locomotory machinery had been left attached to the outside of the TLA, the remnant retained its ability to move. So, the remnant was not just a crumb left dangling on the TLA's mouth, so to speak: it was a zombie husk of a spirochete that could still move but that had been drained of its life.

This combination proved to be a double winner. Internalizing the spirochete's metabolism enabled the TLA-spirochete hybrid to live in more oxygen-rich environments than the TLA alone had been able to do. And by keeping the spirochete's locomotory motors attached to the membrane, the TLA-spirochete hybrid could get to these new environments, powered there by its new locomotory capability—the ability to swim rather than simply ooze through muds—acquired in a manner similar to how the proverbial boy riding a tiger could be said to have "acquired" feline suppleness and speed.

———◆———

New locomotion and new metabolism were not the only things the TLA acquired from the spirochete: it also acquired an entirely new set of tokens of hereditary memory—the spirochete's genes. These carried encoded in them an entirely different remembrance of the past, but when expressed as living memory, they incorporated into the TLA an entirely different vision of how the future should be shaped. This posed a problem for the new alliance, because there is no inherent reason why the spirochete's memory tokens could not have been expressed as insistently as the TLA's own. As each worked to shape the future in their different ways, it is easy to see how inevitable conflicts between them would have arisen. Success of the merger would turn on how these conflicts were resolved.

We know little of how the reconciliation actually occurred—we'll speculate on some possible scenarios momentarily—but the traces of it are evident in one of the eukaryotic cell's more obscure features, the *karyomastigont* (literally "nucleus flagellum," Figure 8.7). This is a system of membranes and tubules that connects the chromosomes and nuclear membrane (the *karyon,* or nucleus) to the basal body and ultimately to the undulipodium (the *mastigont,* literally the "whip"). The karyomastigont is a common feature of eukaryotic cells, but the meaning of it has long been elusive. SET calls it an assemblage of "imperfections and oddities" that are clues to the endosymbiotic origin of the LECA.[21] Some of these represent direct homologies.* The basal body, for example, is supposedly a direct homologue of the assembly of proteins that spirochetes use to attach themselves to another cell. Others are clues to what must have been a spirited contest for dominance between the two sets of hereditary memories. The conversion of the ring-shaped chromonemas of both the TLA and the spirochete into the linear threads of the eukaryotic genome may be an example of this. Snipping unwanted chromosomes apart is one of the common defenses bacteria like *Thermoplasma* deploy to deal with intrusive memory tokens of the bacteria that are their typical prey. The histone complex that plasters the chromosomes of eukaryotes and that controls the epigenetic expression, indeed the definition, of eukaryotic genes appears to be foreshadowed in pro-teins that decorate the chromonema of *Thermoplasma,* and that also are used by them to manage unruly and unwelcome alien memory tokens. The double nuclear membrane of eukaryotes may have its origin as a protective envelope to shield some stretches of DNA from degradation, or to manage how some could be expressed.

We could speculate endlessly on the possible homologies, but

---

* "Homologue" in the sense of a structure deployed to another purpose from an earlier purpose. The wings of birds, for example, are homologues to the fins of fish; both are derived from the same devel-opmental feature of all vertebrates—the pectoral girdle. Homologues are distinct from analogues, which are structures with similar function that have different evolutionary ancestries. The wings of birds are also analogues to the wings of insects, which can be traced back to a modified gill in an arthropod that was ancestral to insects.

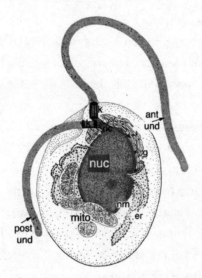

Figure 8.7

Idealized representation of the karyomastigont of a protist. *Post und* and *ant und*, posterior and anterior undulipodia; *nuc*, nucleus; *nm*, nuclear membrane; *mito*, mitochondrion; *k*, kinetosome (equivalent to basal body); *er*, endoplasmic reticulum; *nc*, nuclear connector; *g*, golgi complex. The undulipodium, basal bodies, and nuclear envelope are part of the continuum between undulipodium and nucleus.

there is a deeper and more fundamental homology that interests us: the eukaryotic cytoskeleton itself—that distinctive cobweb of microtubules and microfilaments that permeates the eukaryotic cell. According to some versions of SET, the eukaryotic cytoskeleton in all its diverse forms—centrioles, basal bodies, spindle apparatus, undulipodium—was derived from the flagellar apparatus of spirochetes, which bends and twists the eukaryotic cell into all sorts of fantastic shapes. However, the bacteria themselves are beginning to reveal that they also possess a diverse and complex cytoskeleton.[22] In the eukaryotic cell, the diverse forms of the cytoskeleton can be traced to a ghostly object called a *microtubule organizing center* (MTOC), a self-replicating assemblage of microtubule "seeds" that generate and organize the cytoskeleton in all its multitudinous forms.[23] The sophistication of the bacterial cytoskeleton raises the possibility that bacteria may have MTOCs of some form themselves.

So perhaps it was a spirochete MTOC that was brought into the TLA, there to compete with a TLA cytoskeletal organizer.

No matter what the details prove to be, what is remarkable about the MTOC of eukaryotes is that the MTOC is itself a form of hereditary memory, because its various forms can self-replicate, as nucleic acids do, and in ways seemingly independent of the chromosomes, and in ways that, like all memories, shape the future of the cells wherein they reside. Cells can have one or two or many MTOCs residing within, and they can take various forms. The MTOC also manages the cytoskeleton according to the same strange homeostasis of the dynamic fantasy Mukurob described in Chapter 2, but here carried out at the tiniest of living scales. Finally, the MTOC controls assembly, disassembly, and reorganization of the cytoskeleton according to the physiological demands placed upon the cell.

In short, the MTOC connects the physiology of the cell to the cell's heritable memory in a profound way. The MTOC is the ultimate anchor of the many plot threads we have been following, because its ghostly memory influences everything it touches: the cell's function, its persistence, its adaptability, and ultimately the activation and expression of the multitude of memory tokens within the nucleus. That homology points to something that will allow us to complete our arduous journey between token and memory—and to the surprising conclusion that awaits us there.

———————•———————

Here is the basic problem that the reconciliation of competing bacterial memories had to solve. Both the TLA and the spirochete presumably relied, as all living systems do, on managing an orderly flow of electrons through their cells. We can pick apart many fascinating details of how this flow of electrons powers this or that chemical reaction, and undergraduate textbooks of biology are filled with mind-numbing examples of such things.* The forest this multitude of trees is obscuring is the orderly flow of electrons that is managing an orderly flow of everything else through the cell so that its particular form may persist: homeostasis of the strange variety outlined in Chapter 2. When they were independent, both TLA and spirochete expressed this strange homeostasis in different ways—one in the evanescent form of a spirochete and the other in the evanescent form of a TLA—and each powered its respective flows of electrons in different ways—one by drawing them toward sulfur and the other by drawing them more avidly toward oxygen. So, the putative merger between TLA and spirochete not only brought multiple sets of insistent hereditary memories into conflict, it also brought into conflict two different expressions of homeostasis, as well as two different sources of energy to power the work of homeostasis. Which memory-shaped future would prevail in the merger? Sitting at our vantage point, inhabiting that future, we know the answer: it was the reconciled future vision that gave rise to the LECA. But that is cheap hindsight. So we still must ask: precisely how did the new partners sort out their different future visions?

Modern evolutionism has a ready answer to this question. It was natural selection—Daniel Dennett's "universal acid"—that "decided," which was not really a decision at all, of course, but a kind of filtering of possibilities and propositions. The orderly passage of electrons, energy, and materials through a cell can be managed because genes encode a particular catalytic environment of proteins within the cell. The particular structure of that catalytic environment ensures that electrons, energy, and materials will all flow through the cell in a particular way—a way that bolsters the persistence of the cell's particular form. Nucleotide sequences encode

---

* My favorite (least favorite?) example is forcing students to memorize the details of glycolysis and the Krebs cycle.

amino acid sequences in proteins, which fold up to make enzyme catalysts that collectively either "work"—successfully manage the cell's metabolism—or don't. Those memories that specified catalytic environments that "worked" were able to replicate—were naturally selected—and those that specified inapt catalytic environments did not—they were selected against. Keep this going over multiple generations and the sorting out of hereditary memories that had to have followed the merger of genomes in the LECA would inevitably have followed.

This answer might be superficially satisfying, but it really gets us nowhere because it simply perpetuates the empty tautology of modern Darwinism: what works now works because it worked in the past. There is an alternative way to think about the question, but it is an uncomfortable alternative: what works now works because there is (was) an *intention* to make it work. This is the teleological camel's-nose-in-the-tent that modern Darwinism resists so strenuously, largely because modern Darwinism is philosophically repulsed by the inconvenient rest-of-the-camel: the fraught notions of purposeful agency and desire.

In fairness, it has to be acknowledged that one doesn't need agency to think deeply about life, or that not accounting for agency always leads one into error: anything but! The luminous history of twentieth-century biology is a testimony to the power of that way of thinking. Yet agency seems to be a very large part of what living systems do, and agency seems to extend even down to life at its infinitesimal scale. We should account for it somehow, it seems, but how?

I will make a bald assertion: bacteria (or any living system, for that matter) can be agents because they are cognitive beings. Now, before going any further, I need to insert two disclaimers. The first is that I am using "cognition" in the broadest possible sense I can get away with—to mean simply the mapping of information about the external environment onto the cell's internal workings. The second is that I am distinguishing cognition sharply from consciousness. I feel compelled to do so because the two are often conflated, with the result that neither can be spoken of sensibly. While the two arguably are related to one another, cognition is relatively straightforward compared with consciousness, which is deep, perhaps unfathomably so. It's clarity I'm aiming for here, not profundity, and conflating the two only muddies the waters.

That said, I can confidently say that bacteria can be cognitive agents because they have embedded in their membranes a suite of cognitive mapping tools.[24] These are in the form of protein receptor molecules that respond to environmental conditions and alter the catalytic landscape within the cell. Cognitive mapping of this sort appears to be a universal phenomenon of cellular life.[25] Our own engines of cognition have similar mapping tools: the photosensitive pigments in a rod photoreceptor are membrane-bound proteins that map the presence of photons into an altered physiology within the photoreceptor cell. So, it seems to be cognition all the way down to the simplest life forms we know. Since cognition is an important component of agency, it follows that bacteria can be cognitive agents, as can any living system. There should be nothing controversial about this assertion, as long as we are careful to frame it correctly.

Returning to spirochetes, oxygen is the fuel that powers their peculiar homeostasis. The spirochete doesn't just sit there like a complicated bag of chemistry waiting for the oxygen (or food, or any of a host of other things) to come to it; it actively seeks it, through gleaning information from the environment about where the goodies are and directing its physiology accordingly. Let me rephrase that: the spirochete *intentionally* seeks it. I phrase it this way because, as I have argued in *The Tinkerer's Accomplice*, intentionality is fundamentally the flip side of cognition, and you cannot really have one without the other.* This is because cognitive mapping is invariably connected to engines—cells or organisms—that do work to modify the environment toward a particular—a purposeful—end. In the case of the spirochete, the purposeful end is to sustain the ordered flow of electrons and matter through it so as to sustain the peculiar and evanescent form of the spirochete. This is intentionality at work. Every living system has intentionality, although the intentionality can be manifest in any number of ways. An agent could change its location in space, or it could construct an environment that is more suitable to its physiology. But it happens, inevitably, and this makes life a profoundly intentional phenomenon. That might as well be acknowledged frankly.

This lifts the lid on another, and perhaps more radical, way to think about the merger that gave rise to the eukaryotes. The merger of TLA and spirochete into the LECA brought together many different things: different sets of memory tokens, different forms of homeostasis, and different ways to power the work of homeostasis. But fundamentally, the merger brought together two different cognitive pictures of the world, and two different suites of intentions about how to live in those worlds. Before the merger, the TLA by itself was governed by its own set of intentions, presumably seeking out or constructing environments that were sulfur-rich and oxygen-poor. It could do so because it was cognitively equipped to do so. The spirochete ancestor presumably had different cognitive mapping tools that translated into different intentions, seeking out or constructing environments that were sulfur-poor and at least a little oxygen-rich. If the merger had drawn the spirochete completely into the TLA, that is, as a conventional endosymbiosis would have, the spirochete's cognition and intentionality would have been subsumed

---

* This is what got me in trouble with the reviewer of *The Tinkerer's Accomplice* who was demanding from me a profession of faith. See the Preface.

completely under the TLA's. The completely enveloped spirochete, if it continued to be a cognitive agent at all, would perceive only the environment constructed for it by the TLA. But according to SET, that was not the case: as we have seen, the spirochete's membrane, along with the cognitive mapping tools embedded in it, remained attached to the outside of the TLA's membrane, a zombie husk that not only retained the ability to move, but also the ability to sense and evaluate its environment, and to map its cognitive vision onto the new merged organism.

———•———

This cognitive dimension illuminates the question that began this chapter: why are sensory cells so commonly derived from undulipodia? The reason now seems clear. The ultimate ancestor of the undulipodium is that cognitively-enabled zombie husk of a spirochete. Cognition was thus the undulipodium's function from the get-go. Over evolutionary time, this fundamental cognitive capability has been co-opted and adapted and tinkered into an infinitude of new functions, among them the ability to build astonishing new cognitive worlds. In this sense, the plot threads we have been following have led us back whence we came. This alone would be a satisfying conclusion to our journey.

But we cannot really stop here because this fundamental cognitive dimension has brought to the fore a challenging, and unavoidable, question about the nature of evolution, to wit: what exactly drives it forward? Is it the tokens of memory that force life into an uncertain future, pushing it there to either stand or die? Or is it forward-looking intentionality that strides confidently into the future, dragging the memory tokens along in its wake, *intending* to stand rather than simply to die? The first proposition is the Darwinian idea; the second is the Lamarckian idea. Which do we choose to follow? For much of the twentieth century, evolutionary thought has been dedicated to advancing the first—the Darwinian—by marginalizing the Lamarckian alternative. We are by now familiar with the claim: only one form of hereditary memory matters, and that is the hard inheritance embodied in the gene. No alternative is possible because there is no imaginable way the experience of a life lived can *soften* those hard nuggets of hereditary memory.

It is unimaginable no more, and it is the MTOC that has opened the window on how. The cytoskeleton is not mere scaffolding; it serves as the principal communications network between the cognitive mapping tools of the membrane—the receptors that enable cells to sense the environment—and the physiology of the cell, which includes the expression of the heritable memory.[26] Microtubules position receptors in place on membranes, they hold mitochondria and enzyme complexes in place, and through a collection of mediating proteins called *microtubule-associated proteins,* they connect the cell's cognition to its physiology. Microtubules are the cellular machinery underlying the phenomenon of physiological adaptation.

They are also the means whereby physiological and evolutionary adaptation can finally be unified, for the MTOCs also extend their reach into the nucleus, to the chromosomes and the complex architecture of the chromosome that actually defines the gene. This reach is most evident during mitosis, when the cell's MTOCs bloom into the spectacular spray of the spindle apparatus. It also operates in more mundane and workaday ways to unfold chromosomes, open stretches of them to transcription and expression, and even actively define what genes are. Microtubule-associated proteins also mediate the feedback that closes the loop between soft and hard heritable memory, between physiological and evolutionary adaptation, and between purposefulness and evolution.

The plot threads we have followed through this chapter have now led us onto a very different landscape of evolution. No longer are we stuck in the bleak landscape of the Four Horsemen of the Evocalypse, where there is no purpose, no desire, no intention—only the indifferent churning of a machine. From where we stand now, we can at least begin to see a landscape where those essential attributes of life—purposefulness, striving, desire, intentionality, intelligence—can once again reenchant our understanding of life and of everything about it, including its evolution. In so doing, we can begin to chip away at some of those problems left lying intractable on the purposeless landscape of modern Darwinism.

9

ONE IS THE FRIENDLIEST NUMBER

The first difficult problems we will try to tease apart are individuality and the nature of the organism. These might seem not to be problems at all, simple facts of life, but they are actually problems of perception. We perceive life in ones. The discrete individual is the central fact of our existence and of our relationship to the world. Descartes said it most succinctly: *Cogito ergo sum.* I think, therefore I am.

There are two profound truths in this aphorism. The first is what grabs most peoples' attention—the equivalence of "think" and "am." The second and less obvious truth is the assertion "*I* am." "I" am the central fact of the universe that is "me." To be sure, I look around "me" and see a bounty of other "I's," that is, other individuals who are universes unto themselves, their own existences being the central fact of their lives to them. Nor does it stop with my fellow humans. I have an individual dog, Calloway, and two individual cats, Feral Fawcett and Influenza, each of which are also curious universes unto themselves. Calloway, when he is not sleeping or eating or frantically worrying over his toys, looks at our cats and sees two playthings to challenge and chase. Feral Fawcett and Influenza look at Calloway and see a renegade force of nature that they must, like priests, relentlessly keep in its place with hissy incantations and slashing genuflections. It goes farther still. The wild turkeys, the deer, the occasional coyotes that wander through my property, the fruit trees I cultivate, the Norway maples and sumac that I must constantly keep at bay—all are individuals in their own idiosyncratic ways. Life's ordinal number, it seems, is one: the autonomous, coherent, integrated, adaptable, responsive, intentional, intelligent stream of matter, energy, and information that is wrapped up in the pretty package of the individual.

I am also an organism, and this raises an interesting question: are organisms and individuals the same thing? Am I an organism because I am an individual? Or vice versa? We often seem to conflate the two concepts, as when we speak of the "individual organism." As you delve into the matter, though, the equivalence of individual and organism begins to look a little iffy.[1] Without understanding the relation between the two, how can we point to an evolutionary origin of either? If we define the organism, as we commonly do, as a tangible individual comprising a multitude of genetically related cells functioning together as an autonomous whole, then the organism came onto the scene rather late in the evolution of life on Earth. We see the first glimmerings of it about nine hundred million years ago, finally emerging in full flower only around the beginning of the Cambrian Period, around six hundred million years ago.[2]

This sounds like a long time ago until we reflect that two-thirds of the entire history of life on Earth had already passed by then. Before that, Earth was an exclusively prokaryotic planet, teeming with bacteria. Bacteria are obviously cells, but could we say these were organisms? Cells are autonomous living systems, like organisms are after all; but a swarm of bacteria seems too amorphous to qualify as an organism as we commonly conceive it. Then again, we reflect upon the community of sulfur-breathing bacteria and spirochetes introduced in Chapter 8. In nature, bacteria usually live in layered mat communities, like those ancestors of the eukaryotes did, and this habit seems to have stretched all the way back to the origin of the bacteria,[3] that is to say, near to the origin of life. Microbial mats mimic some of the specialized functions of, say, the organs in the body of a more conventionally defined organism.[4] Could microbial mats be construed as organisms, then? Perhaps they could, if we were willing to stretch the definition. Could microbial mats be individuals? It would be difficult to go that far even by the most Procrustean construction of the term.

The presumed equivalence of organism and individual gets muddier the deeper we look. A striking feature of the evolution of life on Earth is the emergence of a variety of "organism-like" systems, that is, many individuals that coordinate their lives in ways that make them look and behave like organisms. The microbial mat is one example. Another would be "symbiotic organisms" such as lichens, which bring together two, sometimes three, kingdoms of life (typically fungi, algae, and cyanobacteria) into such close association that a lichen looks, behaves, and acts as if it were an individual organism. The foremost example of an organism-like system is, of course, the social insect colony, which, since Théophile de Bordeu, has been likened to an organism that was something more than an individual: a "superorganism," to use the modern term.[5] The social insect swarm exhibits many of the same traits as organisms—coherence, adaptability, collective responsiveness, and even, dare I say it, intentionality.

I'm afraid our confusion is not yet complete. Turning back to look in the mirror, the "I" that seems so unequivocally "me" begins to lose focus as I contemplate that "I" am more of an "us." The assemblage of cells that is my body is populated by ten times as many alien cells—bacteria, yeasts, and fungi—as "my" own cells, and these alien riders carry in them a hundred times more genes than the ones I inherited from my parents.[6] Furthermore, "I" would not function as an "I" or even be the same "I" in their absence. What, then, is "I"? Am "I" an organism, or an organism-like system, or something else? At this point, the long-assumed equivalence of organism and individual becomes very strained indeed.

So, we've worked ourselves again into a bit of a muddle. It's important that the muddle be resolved, because the origin of the multicellular organism—and by implication, the individual—has been tagged as one of the "major evolutionary transitions" in the history of life on Earth.[7] This implies that we know what an organism is, and it seems not at all clear that we do. It also implies that we understand the relationship between organism and individual. Again, can we say honestly that we do? I can't.

Quite a lot is at stake in this question, because an implied individuality sits at the heart of the Darwinian idea. It is individuals that compete with one another for resources, territory, mates; and it is individuals that are either better or worse adapted to the circumstances in which they live. But modern Darwinism has, by and large, pushed that vivid individuality to the wings. The individual is now a mere repository for genes swarming in a disembodied gene pool. The individual is a vehicle, the necessary device for pushing aptfunction genes into the future. Aside from that, the individual as a universe unto itself doesn't really enter into our thinking.

Yet the plain evidence of the senses would reject this view—the individual organism is an insistent fact of nature that it would be hazardous to diminish in our thoughts, even as the beauty of the modern synthesis beckons us to do precisely that. Nonetheless, we cannot resist the siren's song when we are confused about the nature of the organism, its supposed equivalence to the individual, and what has compelled both into being. So we simply have no choice but to pick the issue apart. What are organisms? What are individuals?

How did organisms evolve? How did individuality evolve? Did they evolve simultaneously, or separately to a degree? If there is a common theory to explain the origin of both, would it be Darwinian? Or would it be something else entirely?

———————————◆———————————

There is at hand a ready and long-standing Darwinian explanation for the organism, indeed for the organism-like system. It is called *kin selection theory.*[8] This was originally the brainchild of Charles Darwin (of course), but kin selection's first formal statement came only in the 1920s, summarized pithily by that loquacious leg of the Neo-Darwinian trinity J. B. S. Haldane.[*] When asked whether he would be willing to lay down his life to save a drowning brother, Haldane replied that he probably wouldn't for one brother, but he might do it if he could save two. Or if he could save eight first cousins. The apocryphal version of this story has Haldane scribbling all this out on a napkin while sitting in a Cambridge pub.[†] As David Queller expressed the idea more poetically:[9]

> *Would I jump in a lake*
> *To save my drowning cousin?*
> *It's not a risk I'd take*
> *For him plus half a dozen.*

---

[*] The others being, of course, Ronald Fisher and Sewall Wright. Haldane was the most flamboyant personality of the three, given to pithy quotations and dramatic public gestures too numerous to catalog here.

[†] This is a matter of some dispute. The story implies that someone was in the pub, the Orange Tree, with Haldane when he scribbled out his formulas on that napkin. But no one has ever produced the alleged napkin, nor has anyone come forth as a reliable witness to the event. The closest we get to a confirmation is an acknowledgment from Haldane's sister, Naomi Mitchison, that her brother did in fact frequent that pub and would often scribble out notes to himself on napkins. For a fuller exploration, see Chapter 12 in Ullica Segerstrale's *Nature's Oracle: The Life and Work of W. D. Hamilton* (Oxford University Press, 2013).

But if you raise the stake
And make the prize my brother?
Now that's a deal I'll make . . .
If you'll just toss in another.

Darwin's expression of the idea of kin selection was rather vague, but not so Haldane's, for he had in mind an algebra of altruism that Darwin could not have imagined. Haldane could imagine such a mathematical theory because he could derive it from Mendel's First and Second Laws, laws that were unbeknownst to Darwin. Haldane's algebra of altruism is easy to grasp intuitively, because it brings a gambler's perception of odds to the problem of gene inheritance. Here's how it works.

For any gene in a sexually reproducing species, an individual carries two alleles, one from the mother and one from the father. For any gene sitting within an individual—for argument's sake, let us say Haldane himself—there is a 50 percent probability that an identical copy of the gene will be sitting in his hypothetically drowning brother. This will be true for all the genes that reside within Haldane, because he and his brother had an equal shot at the inheritance of any allele from both their mother and father. The likelihood of sharing genes diminishes with distance of relationship, however. If Haldane had a half-brother (same mother, different father, or vice versa), the probability of sharing identical copies of a gene drops to 25 percent. For his first cousins (same grandmother, different grandfather, or vice versa), the likelihood is halved again, down to 12.5 percent. I could go on, but you get the point.

Being altruistic carries a certain amount of what we might call reproductive risk: altruists put their own reproduction at risk to defend the likelihood that another may survive to reproduce. As any gambler knows, however, risks can be ameliorated by knowing what is at stake and what are the odds. The gene-sharing probabilities that

I've just laid out are the relevant odds when the game of natural selection is played.

This adds an interesting wrinkle to the game that is important to appreciate and can be illustrated by a favorite card game of my own family, Hearts.* Hearts is best played with three or four people, but if more want to play, it's no problem—you just form partners and play the game with two decks of cards. Playing the game with two decks means there is a certain probability that you and your partner will be holding cards with the same face value, say, the Queen of Spades. In Hearts, if you can play the Queen of Spades, you can saddle your opponents with a penalty. With two decks of cards, there is a certain probability that you or your partner will be holding a Queen of Spades, and when playing with partners, it does not matter whether it is you or your partner who plays the card: your opponents reap the same penalty. If your partner can play a Queen of Spades, this will penalize your opponents just as much as it would if you had played a Queen of Spades yourself. You win either way, to the point that it may be good strategy to play "altruistically," that is, to play in a way that accrues penalty points for yourself if you can ensure that your partner can inflict a worse penalty on your opponents by playing the Queen of Spades. The reverse also holds true: your partner may play altruistically if that ensures you can play the Queen of Spades. And if both you and your partner are holding a Queen of Spades, you can both play to inflict substantial penalties on your opponents if you can both play your Queens, smiting your opponents with a double whammy.

Something similar happens in the game of natural selection. In this sense, natural selection would be indifferent to which vehicle, that is, which individual, propels an allele into the future. All that

---

* Hearts is one of the many variants of the basic game of Reversis, which involves accumulating cards of the same suit to form complete flushed sets. All Reversis games involve setting constraints on which cards your opponents are forced to lay down, usually following suit. The rules are readily available on the web.

matters is that the allele is transmitted, and it makes no difference if it is one copy residing in one individual or an identical copy in another individual: the gene is transmitted either way. The odds of a reward are biased by the probability that another individual has the same allele as you. A skilled player of Hearts knows how to estimate the odds so that altruistic play will likely produce an outcome favorable to both partners. Similarly, the odds of relatedness set the odds in the game of natural selection and whether it is a good gamble to be genetically "altruistic," that is, to forgo one's own reproduction to aid the reproduction of another.

As it was left by Darwin and Haldane, kin selection theory was largely a just-so Darwinian explanation for the existence of altruism. With the falling fortunes of Darwinism in the early twentieth century, and the rising fortunes of Thomas Hunt Morgan's mutationist theories of evolution, no one really worried too much that the idea had never been moved from the back-of-the-envelope onto the solid ground of critical test. It was just so darned clever an idea that it had to be true. That all changed in the 1950s, when along came another of those mathematically gifted naturalists in the mold of Fisher and Wright: W. D. Hamilton (Figure 9.1).[10]

William Donald Hamilton (1936–2000) was born in Cairo but grew up in Kent, in what we might describe as an intriguing family. Both his parents were New Zealanders—his father a peripatetic engineer, and his mother a physician. They met one another and fell in love on a ship while both were on their separate ways to take up new lives in England. Once they married, Hamilton's mother gave up her medical aspirations to devote herself fully to her family. Hamilton's father remained an engineer. The backgrounds of both of Hamilton's parents ensured that the household would be intellectually rich.

Figure 9.1
William D Hamilton. 1936-2000.

The Hamilton family was large and self-reliant; the seven children were permitted a great deal of freedom, sometimes to dangerous effect (a foolish experiment gone awry with his father's explosives left Hamilton shy of some fingers and with brass shrapnel permanently embedded in his chest). Mostly, the freedom was salutary. With the family ensconced on its sprawling homestead in the chalk hills of rural Kent, Hamilton grew up immersed deeply in nature.* He also grew to be a brilliant mathematician, but he was a deeper naturalist.

Early in his undergraduate career at Cambridge, Hamilton became obsessed with the problem of the genetics of altruism. How he came to this nicely illustrates the creative, independent, and synthetic way his mind worked. He knew that cooperation among supposed competitors was a widespread phenomenon in the living world. Herds of deer would protect their young, groups of birds would cooperate in

---

* The Hamilton family was also marked by tragedy. One brother died in infancy, and another brother was killed at age nineteen in a mountaineering accident.

the rearing of a brood, prairie dogs would warn others of a patrolling hawk in the sky—all of which were, on the face of it, acts of altruism performed with little immediate benefit for the altruist. Yet there they were, plain for all to see, and to Hamilton, there had to be a good Darwinian explanation for this.

In the Neo-Darwinian world, of course, a "good Darwinian explanation" meant a "good gene-selectionist explanation"; that is, there had to be some theory for how genes for altruism could be selected when the altruistic act put the altruist's own reproductive fitness in jeopardy. Yet aside from the vague kin-selection reasonings of Haldane and others, who called themselves Darwinians but were really Mendelians, a good Darwinian explanation for this had been elusive. Where Darwin and Haldane had basically left the problem was that altruism was in fact a surreptitious form of genetic selfishness: I will give myself up for two brothers, but not for one, because it is only for two brothers that there is a high probability that my genes win in the end, even if I individually end up the loser. The price I will exact for my sacrifice will be higher for my half-sibling, higher still for my first cousins, and so forth. One could, from this impeccable genetic logic, draw the seemingly heartless and cynical conclusion that "real" altruism was a delusion.

Hamilton thought that the problem of altruism had to be deeper than that. His interest in the genetics of altruism came at a bad time, though, because the world had only recently emerged from a bloody encounter with the idea of genetics as destiny—the Nazis had taken this idea, twisted it to support their racist ideology, and run with it, disastrously, leaving millions of corpses in their poisonous wake. For many years following that harsh lesson, few geneticists wanted to hear anything that was tainted with impure thoughts about genetics and behavior, and this made the genetics of altruism the ultimate politically incorrect topic.

So when Hamilton came along wanting to pursue that very idea, he faced enormous hurdles in trying to get anyone in a position of academic authority to even consider it, never mind to contemplate giving an aspiring student free rein to run with it. He was, in Hamilton's own words, looked upon by many of his potential mentors as "a sinister new sucker budding from the recently felled tree of Fascism."[11] It took an extraordinarily clear mind, a willingness to take risks, and an extremely tenacious personality, to follow that idea in the face of such prejudice. Hamilton did, doggedly and brilliantly.

It was not an easy go for him, however. After leaving Cambridge, he decamped to the London School of Economics for graduate study. There, thanks to a few individuals who kept him going with stipends here and there, sometimes a desk to work from, and an occasional behind-the-scenes sympathetic word to journal editors, Hamilton's solution to the genetics of altruism finally made it into the sunlight. Even then, it wasn't all a happy ending: many contemporaries did not understand his solution, or even the problem.

After receiving his doctoral degree, he went on to lecture and do research at Silwood Park, England's premier landing place for entomologists. This seemed to be a good fit for Hamilton's bent for field natural history, but even there, he had a tough go: he was an undistinguished lecturer, was reluctant to publish, and got involved in roiling, sometimes obscure, disputes with his British colleagues. It's fair to say that he would never have gotten far in the bureaucratically dominated modern-day administrative university. He ended up leaving England to pursue a long academic tenure in the United States, first at Harvard and then at the University of Michigan. Only later in his life was he persuaded to return to England, where he took up a Royal Society Research Professorship at Oxford University. In the end, the slights he suffered during his early career didn't matter: Hamilton came to be recognized as the remarkable evolutionary genius he was, with awards and accolades aplenty. Sadly, his life was cut short at age sixty-three from malaria contracted during a field expedition to Congo.*

———————◆———————

Hamilton is best remembered today for what is called the "Hamilton rule."† We shall return to this momentarily, but first we must appreciate some context. Hamilton's famous rule was just one aspect of a larger theory, called *inclusive fitness,* that took Haldane's rather slapdash back-of-the-envelope kin selection theory and put it onto a more sound mathematical foundation. The details of Hamilton's synthesis are too abstruse to go into in any detail here. Suffice it to say that Hamilton reconciled a long-standing theoretical dispute between Ronald Fisher and Sewall Wright over how gene fitness should be measured. Hamilton's synthesis extended the notion of fitness beyond the simple probabilities of gene selection among relatives to encompass all the positive and negative effects of an allele on the fitness of *all* carriers of the allele: the allele's *inclusive* fitness, in a phrase. Hamilton's insight opened the door afresh to the Darwinian analysis of a wide range of social phenomena, including sex ratios, parasite infections, and importantly here, a Darwinian theory of both the social insects and the organism.[12]

This is where we can return to the Hamilton rule, because its most familiar application is to the problem of altruism among the social insects, which include the bees, ants, wasps, and termites. Social insects form complex societies that have taken reproductive altruism to an extreme degree.[13] These insects assemble into large

---

* Ullica Segerstrale's marvelous 2013 biography of Hamilton, *Nature's Oracle: The Life and Work of W. D. Hamilton* (Oxford University Press), outlines Hamilton's colorful life and even more colorful thought.

† The eponymy did not come from Hamilton, who was famously modest and self-effacing, some would say to a fault.

colonies comprising a horde of sterile workers that are the offspring (often) of a single pair of fertile individuals. The sterile workers seem indifferent to their own genetic interests, to the point that they cavalierly sacrifice their lives on behalf of their nest mates, especially their mother, the queen, who is the only fecund member of the colony. This propensity to self-sacrifice caused Darwin enormous theoretical difficulty.[14] Many of the honeybee colony's marvelous adaptations, such as the remarkable hexagonal geometry of the wax honeycomb, were actually the work of sterile worker bees that did not reproduce and presumably could not pass whatever beneficial instincts they had to offspring. Darwin reconciled this difficulty with the Lamarckian dodge that beneficial instincts, such as perfection in comb-building, would become habitual in lineages, provided the instinct was consistently beneficial to the members of the colony that did reproduce.[15] This was Darwin's basic adaptationist approach to the problem, but with the eclipse of classical Darwinism behind the rise of mutationism and gene-selectionism, Darwin's adaptationist thinking ceased to carry much weight. There the conundrum of the honeybees sat, until Hamilton's theory of inclusive fitness brought the problem back for a fresh look through the new lens of population genetics.

Hamilton's rule was his solution, and its application turns on a peculiarity of reproduction among the so-called hymenopteran (membrane-winged) social insects: the bees, ants, and wasps (excluding the termites, which become important momentarily). The peculiarity centers on the determination of sex in the offspring and is best understood by comparison with organisms like ourselves. All sexually reproducing animals are diploid: for each gene they contain one allele inherited from the mother and one from the father. The alleles, meanwhile, come in groups of genes on chromosomes. Whether a child is a boy or girl depends on whether it has a particular sex chromosome. In mammals, sex chromosomes come in two forms, X and Y. Females have two X chromosomes, one inherited from her mother and one inherited from her father. Males, in contrast, have one X chromosome, inherited from the mother, and one Y chromosome, always inherited from the father. At reproduction, the probability that a zygote will receive one X chromosome from the mother is 100 percent,* but there is only a 50 percent probability the zygote will get an X chromosome from the father (and, of course, a 50 percent probability of getting a Y chromosome). In other words, the relatedness of parents to offspring is 50 percent, the same as the relatedness of full siblings to one another. When it comes to the sexes, there will be 50 percent males and 50 percent females.

Not so for the hymenopteran insects: the bees, ants, and wasps. Among them, sex of the offspring is determined not by sex chromosomes, but by whether or not an egg is fertilized. If an egg is fertilized, the diploid offspring is always female. If the egg is unfertilized, the haploid egg contains only those genes inherited from the mother. The haploid egg, when it is left unfertilized, will always develop as a haploid male. Thus, female offspring are always genetically diploid, the recipients of genes from the haploid gametes of both the mother and the father. Male offspring, in contrast, are always haploid: their genome is that of the maternal haploid egg only. This type of sex determination is known as *haplodiploidy*.

Hamilton's rule centers on the implications of haplodiploidy for the inclusive fitness of honeybee colonies. Put simply, haplodiploidy produces families in which sisters are more closely related to one another than they are to their parents, to their brothers, or to any hypothetical offspring they might have if they were to reproduce themselves. Let's put some numbers on this. Among organisms like ourselves, the patterns of relationship are neatly symmetrical. Diploid parents are 50 percent related to their diploid offspring, children are 50 percent

---

* Fifty percent for each of the two homologous X chromosomes.

related to their parents, and full siblings are 50 percent related to one another. In bee families, diploid sisters are 75 percent related to one another, 50 percent related to their diploid parents or to their own potential female offspring, and only 25 percent related to their haploid brothers. Don't worry about calculating these numbers yourself, please just take my word for it. What is important here is that the inclusive fitness of a honeybee family is maximized by sisters turning their mother into their reproductive slave to maximize the production of more sisters. Inclusive fitness is enhanced if that cabal of sisters can work together to glean more resources than other insect societies that might be more, shall we say, sexually normative. And that is pretty much what we see in the honeybee colony: a single reproductive queen and a horde of sterile female workers that exercise tight control over the emergence of fertile reproductive proxies that can mate on behalf of their sterile sisters.

This tight correspondence between social insect reality and theoretical expectation is what cemented the reputation of Hamilton's rule, and it is usually considered to be an explanation for the evolution of the social insects. Hamilton's rule is more extensive than this, though, because it opens the door to explaining other essentially social phenomena, including the emergence of the organism. Hamilton's explanation for this turns on defining the organism in a specific way, that is, as a colossal society of genetically identical cells that we may call the *genetic organism*. Each cell in our bodies is the genetic clone of the diploid zygote formed at our conception. This is not to say, of course, that each cell in the body *functions* identically to every other; one of the hallmarks of the genetic organism is the specialization of the many lineages of descent from the zygote into different types of tissues: muscular, nervous, sensory, excretory, digestive, and so forth. This differentiation comes from silencing some sets of genes and allowing others to be expressed, but the complete genome of the zygote remains present in nearly every cell type in the body.* This means that every cell in the body is 100 percent related to every other, and nearly all of these cells are engaged in an ultimate act of genetic altruism, sacrificing their own ability to reproduce to promote the genetic interests of the small population of reproductive proxies we call the germ line. The explanation for the organism is therefore the same as for the social insect colony, only more emphatically, with many related individuals, the cellular descendants of the zygote in this case, sacrificing their own genetic interests to work together so that reproductive proxies, the gametes, can plant copies of their genes into a new organism on their behalf. It is a beautiful idea—one of the most beautiful in biology, I would say.

Then there are the termites.

The termites are also social insects, but they come from an entirely different order of insects from that of the bees, ants, and wasps. The termites are inconvenient to the beautiful Hamilton's rule because they do not have the supposed genetic bias to sociality that haplodiploidy confers on the hymenopteran insects.[16] The sex of termites is determined in the ordinary way, by sex chromosomes, so the female bias that is prevalent among the hymenopterans is absent from the termites. Because termites are not haplodiploid, patterns of relatedness in termite families are as neatly symmetrical as they are in our own families: uniformly diploid offspring that are 50 percent related to parents, to their siblings, and to their own potential offspring. Yet the termites build complex insect societies that are remarkably convergent on the wonderful societies of the bees, ants, and wasps. A termite colony consists of hordes of sterile

---

* Save for a few specialized types like the red blood cells of mammals, which have expelled their nuclei altogether.

workers turning their parents into their reproductive slaves* and working tirelessly to advance the reproductive interests of a small cadre of reproductively privileged proxies. Termite societies are also complex, with specialization of tasks among the workers, just as one finds in ant and bee colonies. The lumpen termites stand up respectably, it seems, against the more glamorous bees, ants, and wasps, despite not conforming to the genetic biases toward sociality that Hamilton's rule would imply.[17]

Hamilton was well aware of this problem and proposed an ingenious solution to it.[18] He thought the genetic bias to sociality need not come from haplodiploidy alone; it could also come from a high degree of inbreeding, which should be common among the termites. The inbreeding itself might have been secondary to another trait of termite societies: their reliance on intestinal microbes to digest their woody diet for them.[19] Termites cannot live without these gut microbes, but they lose these microbes with each molt. To continue to digest wood, termites' intestines need to be continually reinoculated with cultures of these microbes.† The most reliable source of inoculum would be other termites, specifically the feces of other termites, and getting access to the inoculum is enhanced by having lots of other termites close by: sociality, in a word. Enforced sociality like this would promote incestuous breeding, particularly if the insects were sexually libertine—to paraphrase one of the more cringe-worthy mottos to come out of

---

* Sometimes literally. Among some of the more advanced termites, the queen and her consort are imprisoned in a worker-constructed queen cell, where she spends the nearly two decades of her life as an egg-laying machine, turning out roughly twenty-six thousand eggs per day.

† Akin to how we eat active culture yogurt to reinoculate our own intestines following a round of microbe-killing antibiotic therapy. This is also the same logic behind the emerging and promising therapy of the misnamed "fecal transplant" for treating long-term immune disorders of the colon, such as Crohn's disease or irritable bowel syndrome. Here, the "fecal transplant" is actually a "microbiome" transplant, introducing into one person's disordered colon a new microbial community that could restore the colon "ecosystem" to normal function.

1970s popular music, loving the ones they're with.*

Now, we normally think of inbreeding as a bad thing, and with good reason, but from an inclusive fitness standpoint, inbreeding would have the perverse effect of increasing the sterile workers' inclusive fitness. By increasing the relatedness among all the members of a colony above the theoretical 50 percent, inbreeding would make reproductive altruism a less risky proposition. To dredge up Haldane's algebra of altruism again, Haldane might well have taken the risk to save only one brother as long as the brother was his identical twin, that is, someone who shared 100 percent of his alleles rather than the 50 percent an ordinary brother might.

It's all very neat and beautiful—the organism and the social insect superorganism explained in one pretty package wrapped up with a bow of inclusive fitness. Unfortunately, nature, as is often the case, is no respecter of intellectual beauty. Consider, for example, the haplodiploid bias to sociality among the hymenopteran insects. For this to work, mates must be faithful—one diploid queen mating with one haploid male. But this seems rather not to be the case. Bee queens in nature are not beacons of marital virtue but are ravenous couplers that would put to shame the emperor Claudius's notorious consort Messalina.† As a consequence, the sisterhood that issued from these

---

* The line comes from the lead track on Stephen Stills's 1970 eponymous solo album. The song was said to be inspired by an offhand remark by Stills' contemporary, Billy Preston, and reflected the free-wheeling sexual ethos of the time. The sentiment of the song has come under criticism as reflecting a casual acceptance of adultery and infidelity, not just from the culturally conservative, but from his fellow musicians. For an unquotable example of the latter, see David Yow's interview at the A.V. Club online, "David Yow on Why He Hates 'Love The One You're With,'" www.avclub.com/article/david-yow-on-why-he-hates-love-the-one-youre-with-99481.

† The third wife of the Roman emperor Claudius, Messalina had a reputation for rampant promiscuity. She was clandestinely executed for her role in a treasonous plot against Claudius that was led by one of her many lovers, the senator Gaius Silius. Her bad reputation, cemented into place several decades after her execution, was mostly the work of Pliny the Elder, along with histories by Tacitus and Suetonius, which portrayed her as a ruthless and sexually insatiable monster. This smearing of Messalina may have been motivated in part by a rebellion against the Julio-Claudian dynasty of emperors, to which Messalina had belonged. The Julio-Claudian dynasty came to an end with the suicide of Nero in 68 CE.

promiscuous matings of their mother with multiple males would not be as closely related as they would have been if the queen had been monogamous. And, by the logic of Hamilton's rule, less likely to be altruistic. The bee sisterhood may be more like *Mean Girls* than a powerful cabal of cooperating Amazons.*

It's among the termites that other, more telling, anomalies start to show. Termites have a variety of social and mating systems among them, and these provide an interesting test of the idea that inclusive fitness is a motivator of sociality, and by implication, of the organism (and of the superorganism).[20] Among the subterranean termites (*Reticulitermes*) of North America, for example, family relationships are very fluid. Colonies are dispersed and subcolonies coalesce wherever there is food and shelter. This social fluidity is one of the reasons these termites are such dangerous pests of wooden houses in North America. What is most interesting about them is the workers' rather loose commitment to their own sterility. While *Reticulitermes* workers usually remain sterile throughout their lives, they can revert under some conditions to a quasi-larval state and become fertile with their next molt. The result is frequent "reproductive break-outs," where workers decide to quit behaving altruistically and to advance their own genetic interests selfishly. "Selfish" in this context means a high rate of incestuous matings among *Reticulitermes* workers that, according to the logic of Hamilton's rule, should increase their inclusive fitness.[21] *Reticulitermes* colonies, by this logic, should evolve toward more rigid societies, but just the opposite seems to be the case: *Reticulitermes* societies are very fluid.

Standing in contrast to this are the colonies of the advanced fungus-growing termites of sub-Saharan Africa and southern Asia (*Macrotermes*). These live in consolidated subterranean nests,

---

* *Mean Girls* was Mark Waters's comedic 2004 adaptation of Rosalind Wiseman's book *Queen Bees and Wannabees* (screenplay co-written by Tina Fey). The story focused on destructive jealousies among teenage girls in high school and building communities of trust rather than competition.

concentrating one or two million workers into a highly structured "termite city" that is roughly 2 meters in diameter.[22] The queen lives for ten to twenty years, and she mates for life with one male (*Macrotermes* represent the sole case of which I am aware of monogamy among the insects). This means there is little opportunity in a *Macrotermes* colony for the incestuous hanky-panky that, according to Hamilton's rule, would boost the colony's inclusive fitness.

These contrasting social systems prompt a question, which immediately leads us into a paradox: which of these termite societies produces the most "organism-like" superorganism? I realize I am asking you to compare degrees of "organism-ness," which I would agree is a rather nebulous concept. Bear with me, though, because we do form rather strong impressions of what organisms are, just as Justice Potter Stewart could form a strong impression of obscenity, even if he could not define it objectively.* If inclusive fitness drives the evolution of the organism, we would predict that the more incestuous *Reticulitermes* colonies would be more "organism-like" than the more sexually normative *Macrotermes* colonies. In fact, however, it is just the opposite. *Reticulitermes* colonies are diffuse blobs, spreading through forests like mold spreads across an old loaf of bread. The individuals are loosely held in association with one another, with renegade matings cropping up everywhere. In contrast, *Macrotermes* colonies are compact, as our own organismal bodies are; and there is strong specialization and differentiation of labor, structure, and function, just as is the case for the multitudinous cells of our own bodies. The impression of "organism-ness" is compellingly strong for *Macrotermes.* For *Reticulitermes,* not so much. If it is inbreeding that drives termite sociality, Hamilton's rule would predict the opposite.

---

* Justice Potter Stewart, in the landmark case *Jacobellis v. Ohio* (1964), which tested laws for banning pornography, expressed in his opinion: "I shall not today attempt further to define the kinds of material I understand to be embraced within that shorthand description [hard-core pornography], and perhaps I could never succeed in intelligibly doing so. But I know it when I see it, and the motion picture involved in this case is not that."

Just to drive a final shiv into the ribs of the Hamilton rule, symbionts confuse the meaning of inclusive fitness totally, for the question now becomes, Who is advancing whose inclusive fitness? The cultivation of fungal symbionts by *Macrotermes* means that not just one, but many genomes are at play, all of which depend vitally upon one another for their collective continued existence. At this point, the notion of the genetic organism—a collection of genetically related entities working together to advance a commonly held genetic legacy—dissolves into smoke. The problem is simply compounded if, as it's beginning to look, we are all symbiotic organisms, the carriers of not one, but multiple and diverse genetic legacies.

———•———

So, what's wrong? Here's my answer. Inclusive fitness is a theory of the genetic organism, and so it is a legacy of the trap of Darwinism as gene selection. If Hamilton's theory of inclusive fitness leaves important questions unanswered or unexplained, as I have argued that it does, where does the problem of the evolution of the organism lie? I'm personally disinclined to blame the theory of inclusive fitness—it is a brilliant idea—but if that is not where the problem lies, perhaps it is in its predicate: the premise of the genetic organism.

The idea of the genetic organism, like the species concept itself, is recast from an older conception of the organism whose philosophical roots do not sit comfortably in the materialist metaphysics of modern biology. That older conception embodies in the organism an ideal of autonomy, integration, purposefulness, and intentionality. In this conception, the organism is life's unique expression, and this makes the organism a quintessentially vitalist idea.

This should not surprise us by now. As we have seen, it was vitalism (of the scientific variety) that animated much of the life sciences of the nineteenth century. Physiologists like Claude Bernard were concerned with questions about how the complex organism works. Embryologists like Brooks, Driesch, and Morgan (when he was an embryologist) were concerned with the question of how the complex organism arises from its simple beginnings as an amorphous zygote. Evolutionists like Lamarck, Cuvier, Cope, and to a conflicted extent Darwin were concerned with how new species, that is, new organismal forms, evolved.

Toward the end of the nineteenth century, the realization began to grow that the three ideas —function, development, and evolution— had to be related to one another somehow. The most exciting intellectual ferment centered on the relationship between evolution and embryological development, which mixed together into a bubbly froth as vitalist thought competed with biology's emerging materialism. With biology's turning away from vitalism, a new concept of the organism was needed, so the organism concept got a makeover. It was bathed, dressed up in new clothes, rouged and coiffed, and cleansed of its vitalist odor so that the organism could now mingle, like Eliza Doolittle at Ascot, with the polite society of modern mechanistic biology.* Thus was the genetic organism born.

The makeover proved to be very fruitful. The emergence of a complex organism from the simple beginnings of a zygote had been a mystery for centuries, with opinion (for that is all it could be) revolving around two possibilities (Figure 9.2). Perhaps there was some preexisting form in the egg that unfolded during development to produce the complex organism ("preformationism"). Or perhaps the complex organism arose from some organizing force that imposed

---

* From *My Fair Lady*, George Cukor's 1964 film adaptation of George Bernard Shaw's 1913 play *Pygmalion*. In Shaw's play, Eliza Doolittle is a Cockney flower girl who is taken in by phonetics professor Henry Higgins, who trains Eliza to speak in a "proper" British accent. Shaw's point was to illustrate how language reinforces class distinctions. The initial test of Higgins's idea was to present Eliza at the annual Ascot races to see if she could pass as upper class. In the film, the luminous Audrey Hepburn portrays Eliza at Ascot with hilarious effect. See the YouTube clip "Eliza Blunders the Small Talk," www.youtube.com/watch?v=8uozGujfdS0.

a.

b.

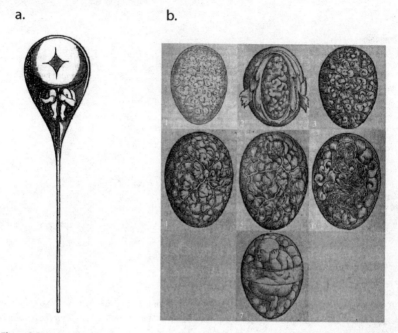

Figure 9.2

The essential conflict of embryogenesis. *(a)* Preformationism, illustrated by the homunculus in the sperm. The complex organism was the unfolding of an already existing form in the gametes. *(b)* Epigenesism, where form emerges spontaneously from the unformed zygote, guided by a vital force. Neither was entirely adequate.

order on an amorphous zygote ("epigenesism"). Centuries of philosophizing had provided no real resolution, but the twentieth-century geneticists thought they had the tools finally to cut through the philosophical muddle.

By identifying genes that specified development, geneticists could now train a clear objective lens on the vexing question of how the complex form of the organism arose. This had modest beginnings as Morgan and others compiled lists of mutations that affected particular body traits of the fruit fly—eye color, wing length, number of segments in the body, and so forth. The culmination of this merger of genetic specification with embryonic development has been the gene regulatory network, an example of which is shown

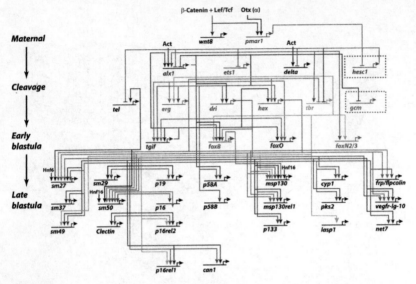

Figure 9.3

The gene regulatory network concept of the genetic organism. The complex organism is the result of a sequence of activation and de-activation of developmental genes, shown here from the maternal stage through to the blastula stage, where the basic architecture of the gut has become set. Note the increasing complexity of gene activation (arrows) and gene suppression (bars).

in Figure 9.3. The details of this particular example (it is concerned with the origin of the mineralized tissues of sea urchins) should not cloud the larger idea this type of thinking exemplifies: that embryological development is essentially a genetic algorithm.

It's a powerful metaphor, to be sure. Algorithms play out through timed and specified sequences of operations, tantamount to switching components off and on. In the case of a computer's central processing unit, the components are the maze of transistors and logic gates etched onto a microchip. Different patterns of switching can make a CPU do interesting and valuable things, from adding two numbers together at lightning speed to predicting the course of hurricanes to winning at *Jeopardy* and everything in between. To do any of this successfully, there has to be a requisite sequence and pattern of switching operations, and the algorithm ensures it will come off properly. In the case of a developmental algorithm, genes specify when and where other genes might be activated (switched on) or deactivated (switched off). These might in turn activate entirely different genetic subalgorithms that manage particular developmental events, such as the emergence of an eye or a limb. Because body form is the product of these developmental events, evolution of body form then becomes the evolution of the genetic algorithms that control development.

The genetic organism is therefore a euphemism for the organism as algorithm. Life is code. Evolution is modification of code. We are all beta versions of something, with infinite updates coming. The question nags, though: code for what? Anyone who has done any coding appreciates that an algorithm must *do* something, and that something usually begins as a desire somewhere in the mind of a coder. The genetic theory of natural selection is supposed to have done away with the need for a coder and his messy intentions, because good codes will naturally survive and replicate, while bad codes will disappear. That's not really a satisfactory solution, though, because it drives us right back to the tautology of modern Darwinism: the well-adapted organism is the code that specifies the well-adapted organism. So, we're again left stuck in a muddle.

———— • ————

Here is where homeostasis might provide a lifeline. Claude Bernard conceived of the organism as a well-regulated internal environment that is demarcated from an ambient environment. All the physiological bells and whistles that an organism comprises—all those delicious and fascinating functional details—are aimed toward a single end: maintaining the constancy of that internal environment and thereby the persistent existence of the organism.

In this conception, the organism is not code, and neither is it an assemblage of code tokens composing the genome. Rather, the organism is *process,* specifically the process of homeostasis. That process might take different forms, but in all instances, a successful process is one that sustains the form most reliably in the widest possible range of environments. In short, homeostasis is persistence and persistence is fitness. This is quite a different conception of the organism, but it must be said that it is not a particularly new one: it is Claude Bernard's conception, indeed the vitalist conception, of the organism. It is an intriguing conception, because it allows us a new way to explore the relationship between the individual—ultimately the agent of evolution—and the organism. Among the things we will discover is that Bernard did not take his conception of the organism to its logical end.

Bernard's conception of the organism was fundamentally a theory of environments partitioned into living (*milieu intérieur*) and nonliving (ambient environment). What separates the two is something we will call an *adaptive interface,* which manages the physiological bells and whistles that underlie homeostasis. At its most fundamental level, the cell membrane can be an adaptive interface, but adaptive interfaces exist at many different scales: assemblages of cells, organs of the body, the organism itself. No matter what the scale, an adaptive interface comprises the numerous machines that do work to power the flux of materials across the interface, that is, between the *milieu intérieur* and the ambient environment. Homeostasis is the outcome when these material and energy fluxes are managed in a way that sustains the internal environment, even if the ambient conditions change. Homeostasis therefore boils down to what happens at this interface, or more precisely what happens *across* the interface.[*] Let me illustrate with a simple example where the adaptive interface is the cell membrane.

Cells closely manage the salt content of their internal environments

---

[*] I've written about this more extensively in *The Extended Organism: The Physiology of Animal-Built Structures* (Harvard University Press, 2000).

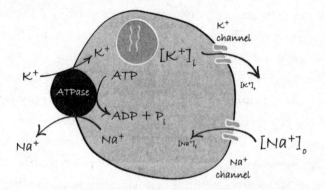

Figure 9.4
The managed interior environment of the cell as regards sodium (Na) and potassium (K) content.
Details in text.

(Figure 9.4), in particular, the concentrations of sodium chloride, or common table salt, and potassium chloride, the common "salt substitute" for people who must restrict sodium intake. Within a cell, that is, in a cell's *milieu intérieur,* potassium is abundant, but sodium is sparse. The opposite conditions prevail in the cell's ambient environment: sodium is abundant, while potassium is sparse. Large differences in concentration of the two salts therefore exist across the cell membrane, made possible because machines in the cell membrane, called sodium-potassium ATPases, do work to drive sodium across the membrane out of the cell while simultaneously drawing potassium in.

The activity of these so-called ion pumps is regulated together with ion "channels" that allow both sodium and potassium to diffuse across the cell membrane—sodium leaking into the cell, and potassium leaking out.* You can think of the cell, therefore, as a leaky bucket. To keep water in a leaky bucket, you have to pump water in as fast as it leaks out. Similarly, high potassium levels within

---

* The distribution of ions like sodium or potassium across the membrane is complicated by the fact that the ions carry charge, and this can affect how these two ions move. This is governed by a small voltage that typically occurs across the membrane.

the cell can be sustained only because work is being done to pump potassium in across the membrane at a rate sufficient to offset potassium's leakage out through the potassium channel.

Homeostasis is therefore a balancing act between ion pumps and channels that, if functioning together, produces a steady composition of the cell's *milieu intérieur,* even as the salts are flowing continuously through the cell, and even if the salt composition of the external environment changes. The adaptive interface of the cell membrane is the pivot point of this balancing act. The interface is adaptive because it gauges the work it must do by the demands of an uncertain ambient environment.

So far, this has been a conventional description of homeostasis, an operational definition Claude Bernard would have been quite comfortable with. It's important to note, though, that this is an *intensive* definition of homeostasis, that is to say, it is focused on the *milieu intérieur,* which is only one side of the cell's adaptive interface. What Bernard did not see was that physiology must also be *extensive;* that is, any modification of the cell interior necessarily affects the environment *outside* the cell as well.[23] There is no magic at work here; it is a straightforward consequence of the principle of conservation of mass. Every potassium ion pumped into the cell depletes the external environment of one potassium ion. Every sodium ion pumped out of the cell enriches the external environment by one sodium ion. Therefore, homeostasis is necessarily both *intensive* and *extensive*: it can be no other way. Homeostasis, indeed all physiology, cannot be confined to an internal environment, as Bernard implied; all homeostasis must be *extended* homeostasis that encompasses both internal and external environments.

Extended homeostasis might be true de jure, so to speak. It follows from a true premise—the conservation of mass—so if we accept the premise, the logical argument that follows from it also must be true. It is fair to ask, though, whether extended homeostasis should ever be de facto relevant to anyone's thinking. Here's why the question is important. It's fine to argue from first principles that something exists, but its importance to the "real world" may be minuscule. When a mischievous little girl dips a bucket of water from the ocean to pour it over her unsuspecting friend's head, there definitely will be a drop in the sea level, but it will be so small as to be utterly irrelevant. Similarly, a single cell floating in the Pacific Ocean and pumping potassium into itself will not appreciably change the ocean's concentrations of potassium. There's just too much potassium there for the cell's internal homeostasis to make much of a dent in its external environment. There's also the fact that external environments are buffeted and disrupted by vast physical forces that swamp life's puny efforts to control them. In such circumstances, the extended homeostasis of a single cell, no matter how logically sound it must be, will matter about as much as the draft of a gnat's wing against a hurricane.

Even so, the purported extended homeostasis cannot be ignored. Remember that extended homeostasis is not some half-baked phantasm; it is derived from the rock-solid physical principle of conservation of mass. Even if the effects of some purported extended homeostasis on the external environment might be negligible, the phenomenon still will have some impact, somewhere. The impact could be on the work of homeostasis, which the cell itself must pay. Energy must be expended to pump that potassium in against the high concentration already there. That energy cost will vary depending upon the potassium concentration outside the cell, and that might vary in both magnitude and time. This means that homeostasis imposes energy costs that are driven by the inhospitability or unruliness of the external environment. To the cell, such costs are not negligible, even if the effects on the external environment might be negligible. If those costs are large or unpredictable, this may mean that less energy is available for the cell's reproduction or persistence.

Some fitness premium should therefore accrue whenever those costs can be reduced or otherwise brought better under the cell's control.

What, then, is a cell to do if the environment changes, or somehow imposes higher costs for homeostasis than the cell is capable of mobilizing? More to the point, what is a cell *lineage* to do if it is to persist? The Neo-Darwinist option would rely on gene selection, that is, rely on natural selection to favorably alter the genetic specifications for the machinery of homeostasis—more efficient ion pumps, tighter ion channels, that sort of thing. This could be pretty quick, provided enough latent variation existed in the genome to extend the range of apt function. Or, it could take a very long time, waiting for a mutation of some sort to provide a new part for the homeostasis engine. The latter case, the Neo-Darwinist case, is very wasteful, because there will be many "guesses" (mutations) for what the "right" part will be. If a guess turns out to be wrong, that's the end of the game for that lineage.

There is a faster and easier solution, though, which life seems to have seized upon quite often: internalize unruly external environments behind new adaptive interfaces. The kidney provides a useful example for how this works. The kidneys manage the water and salt composition of the body's extracellular fluids, in which the body's innumerable cells bathe. The kidneys are made up of a complex suite of tubules, which themselves are made up of sheets of cells folded into the tubules (Figure 9.5). This sheet of cells, called an *epithelium*, comprises a new adaptive boundary. The kidneys pull off the trick of managing the salt and water content of the body by managing the flux of salts and water across the tubular epithelium.

Now let us ask: what in this example is *milieu intérieur* and what is ambient environment? The answer is not self-evident. To Bernard, the *milieu intérieur* would be the extracellular fluid. But the cells of the body would beg to differ: to them, the *milieu intérieur* is the cytoplasm and the extracellular fluid is the ambient environment

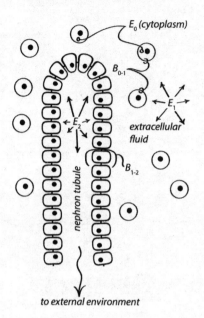

Figure 9.5

Organization of the kidney (nephron) tubule. The tubule epithelium is an adaptive interface (B_{1-2}) between two environments: the extracellular fluid (E_1) and the interior of the tubule (E_2). The cell membranes of the body's cells are adaptive interfaces (B_{0-1}) between the environment of the cell interior (E_0). The interior of the tubule is contiguous with the ambient environment.

(Figure 9.5). The cells of the kidney tubule, for their part, agree with Bernard. To them, the *milieu intérieur* is the extracellular fluid, and the ambient environment is the fluid contained inside the tubule. From the perspective of the kidney as a whole, though, the *milieu intérieur* is the fluid within the tubule. Even though the interior of the kidney tubule is topologically external to the body, that is, it is contiguous with the ambient environment, what comes out of the tubule at the urethra is quite different from what initially shows up in the tubules. This makes the kidney itself an adaptive interface, and therefore an engine of homeostasis.

I think you can now see the point: which is *milieu intérieur* and which is ambient environment is defined by which of several nested adaptive interfaces we are talking about (Figure 9.6). The costs of

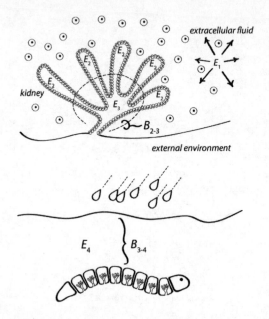

Figure 9.6
Nested adaptive interfaces. The complex kidney, which includes the tubules, constitutes a new adaptive boundary (B_{2-3}) that manages flows of matter between a new environment (E_3) and the tubule environment (E_2). Animals can create new adaptive boundaries in a constructed external environment (E_4), exemplified by the earthworm turning the soil into a new adaptive boundary ($_{B3-4}$) that manages availability and flux of water.

homeostasis that burden the cells of the body are now borne largely by the cells in the tubule epithelium, which have enfolded the tubule's external environment into an internalized environment. The costs of homeostasis for the tubule epithelium, in turn, are eased by the kidney folding the tubule's "external" environment into a new internalized environment of the kidney. Thus, the tubule represents the extended homeostasis of the cell. The kidney represents the extended homeostasis of the tubule. What comes out of all this serial nesting of adaptive interfaces and internalized environments is the organism. The organism is the sum total of the extended homeostasis of its subsidiary parts.

The logic of this "extended organism" needn't stop at the organism conventionally defined. Earthworms, for example, modify soils to provide themselves with the essentially aquatic environments for which their own kidneys suit them (see Figure 9.6). In this instance, the modified soil is part of the earthworm extended organism: it is an adaptive interface between the harsh desiccating environment and the earthworm's essentially aquatic kidneys. Similarly, social insect colonies build nests, some of them of stunning complexity, that function as adaptive interfaces between a now internalized nest environment and the external environment. The logic of the extended organism extends, it seems, to all organism-like systems. Organisms at any scale—cell, epithelium, organism—are properly *extended* organisms, continually working to draw the environment into a conspiracy of extended homeostasis. The conspiracy is vast and diffuse, and it is limited in its reach only by the confines of the biosphere itself.

The extended organism concept effectively treats the organism as a focus of homeostasis rather than as a collection of memory tokens. The individual that is "me" need no longer be limited to the lineal descendants of my zygote: it can also include all the microbial riders that populate my body. All contribute to the physiological conspiracy that is the extended homeostatic "me."

The extended organism idea also seems to dissolve whatever equivalence there might be between organism and individual. "I" am no longer an autonomous being but a superorganism. "I" includes the multitude of genetically diverse microbial riders, as well as numerous other adaptive interfaces that extend outward from "me": the farmers who grow my food, the soil microbes that help the farmers, the government bureaucrats who ensure that the little pieces of paper I give the farmers will compensate them fairly for their efforts, the contributions of the vast assemblage of taxpayers who give me the little sheets of money in the first place. In this tangled web of physiological conspiracies, the individual that is "me" dissolves away into the vast collective "we."

At this point, we might conclude that the individual is as dead an idea as Nietzsche famously pronounced God to be. "I" am but a figment of my own imagination. There is only the vast extended organism; there is no individuality. But this conclusion is unsatisfying on many levels, not least because it conflicts with what seems to be another insistent fact of nature: the individual *exists*. I *know* I am an individual. That is as incontrovertible a fact to me as the ground upon which I stand. If we are led to a conclusion that seems to negate this fact, either the argument itself is wrong, or there is more of the argument yet to follow.

There is more, and the "more" of the argument is homeostasis, properly understood, which leads us to a new conception of the organism, as well as a new conception of the relationship between individual and organism. Let us return to the "Mukurob model" of homeostasis, outlined in Chapter 2: mass flow (sand into the Mukurob) is coupled to energy flow (the work needed to organize the inflowing sand) to produce a specified dynamic disequilibrium (the ephemeral Mukurob). The mass and energy flow parts of this definition are covered by well-established physical principles, like conservation of energy and mass. The "specified" part is more problematic, because it prompts the question "specified by what?" Modern evolutionary biology is there with *its* ready answer: the orderliness is specified by genes, codes, algorithms, whatever, that have been naturally selected. This leads us right back to the concept of the genetic organism, with all its attendant problems and tautologies. Among them is the negation of the individual as an agent of evolutionary adaptation. In this conception, organisms evolve, but individuals do not. Conversely, individuals may adapt, but organisms do not, at least not in the same way. The individual is thus an illusion; only the organism is meaningful. The organism and individual are cleaved asunder.

There is a way out of this unsatisfying logic, however, and it involves cognition. The extended organism, defined as it is as a focus of homeostasis, is actually a *cognitive* organism, cognitive in the same sense that the coalition of sulfur-breathing bacteria and spirochetes from the previous chapter constituted a cognitive entity. Homeostasis involves coupling information about the state of the environment on one side of an adaptive boundary to the matter and energy flows across the adaptive boundary. Now the notion of what individuality is becomes clearer: the individual is a cognitive being that has a sense of itself as something distinct from its environment.

The sense of individuality need not be a *conscious* sense of self. I am consciously aware of myself as an individual, but I presumably could not be so without there first being a *cognitive* sense of myself. Nor should there be any reason to suppose a cognitive sense of self need be coupled to a conscious sense of self. It is reasonable to argue that cognition is a property of all living systems, including the physiological conspiracy that is Gaia—the notion that the biosphere is a colossal living organism.[24] It's probably unreasonable to speak, as some of the more far-out advocates of Gaia theory sometimes do, of the Earth's "global consciousness"; but it's not so unreasonable to speak of there being a "global cognition." Finally, individuality is liberated from genetic lineage. I am not an individual because of the unique combination of memory tokens I inherited from my parents. Rather, I am an individual because I am cognitively aware of that fact, even if I'm not always consciously aware of it. My microbial riders, for example, can modify my cognitive sense of the world and can modify it so that I (we) interact with it differently, ranging from what foods I strive to eat, to how I think about my body, to whether my cognitive sense of the world conforms to the "real" world that is out there.

What conclusions may we now draw about evolution and adaptation? The play of natural selection is a play of individual characters. Lots of behind-the-scenes staff may be fiddling with the sets, lights, and curtains, but the competition and striving for survival that is natural selection is a play of the individual actors on the ecological stage. When we define those individuals as cognitive entities, we are also defining them as *intentional* entities. And here is our startling conclusion: evolution now becomes a phenomenon driven largely by the intentions of the cognitively individual actors. It is not the memory tokens of the genetic organism that thrusts life into the future, but the other way round. It is the striving of cognitive individuals that reach into the future and drag the memory tokens along in their wake. Perhaps most important, the organism and individual are reunited into a cognitive whole that both lives with desire and evolves with purpose.

Let us take a moment to contemplate that last sentence. I have taken you on a journey starting with Claude Bernard's conception of homeostasis as a fundamental property of life: a quintessentially vitalist idea. We saw how this vitalist philosophy permeated early speculations about the evolution of organisms, as expressed by Lamarck, Cuvier, and to an interesting extent, Darwin himself. We saw how this vitalist idea was gradually squeezed out of biology by the pursuit of life as mechanism, devoid of the attributes that make life a distinctive phenomenon in the universe. This pursuit produced such a formidable intellectual landscape that it seemed impossible to challenge. I have tried to show that such a challenge is not only possible, but when looked at through a framework of homeostasis, is eminently plausible. In doing this, we can now reconnect evolution with physiology, resurrecting Lamarck's original vision of a unified theory of adaptation. That vision encompasses both organisms and lineages and makes evolution a living phenomenon once more, marked by intentionality, purpose, and desire—the very things we have, for decades, been warned to shun.

There is another large question looming. How did life come to be? The Darwinian idea has no good answer for this. Neither does anyone else, by the way, but that should not stop us from asking whether, in light of our new way of thinking about evolution, driven by purpose and desire, at least a plausible explanation is possible.

10

THE HAND OF WHATEVER

In 2010, the news was abuzz with the latest claim that life had been created in a test tube. She was nicknamed Synthia, the first "synthetic organism," and she was the fulfillment of a long-standing goal of the J. Craig Venter Institute (JCVI). The media were filled with predictable hyperbole—stories of brilliant scientists "creating life" and "playing God," oblivious to the dire warnings of the existential threat that artificial life posed to us all.[1] People were on the edges of their seats wondering what the pope would say (the Vatican was fine with it). Academic jealousies and snark fulminated below the surface. The JCVI helpfully added to the buzz by putting out an hour-long infomercial titled "Synthesizing Life" that was airing on the BBC within a week of the announcement. It obviously had been prepared well in advance. Craig Venter himself boasted that Synthia was the first organism ever to have a computer as its parents. Synthia was even given a faux Linnaean species name: *Mycoplasma laboratorium*.*

Hype aside, Synthia was a remarkable achievement: she was, in

---

* The species name *Mycoplasma laboratorium* does not appear to be officially recognized by the international body that governs the recognized species of bacteria. See LPSN bacterio.net, List no. 151, http://www.bacterio.net/-validationlists.html#centcinquantetun.

fact, a scientific tour de force. One thing Synthia was not, however, was synthetic life, and here is why. Bringing Synthia into being involved first sequencing the single chromosome of a mycoplasm, probably the simplest known bacterium (although it is still very complex, let me add). Venter's group then synthesized snippets of the mycoplasm chromosome using a machine that could produce short strings of DNA with a specified order of nucleotides. That is where the synthetic part ended, for what followed was a series of ingenious steps that needed naturally existing life forms to be performed. First, the snippets had to be stitched together into something resembling an actual chromosome. Living yeast cells helpfully provided that service. These new chromosomes were then injected into living bacteria, which were able to access and express genes from the synthesized chromosome. Because these crucial steps required already living cells, it would have been more accurate to call Synthia "life-facilitated synthesis." It was simple overreach (although probably an outstanding way to sell newspapers and to raise prestige and more capital)* to claim that Synthia was "synthetic life."

Synthia's story is important because it taps into deep-seated myths of hubris and nemesis, themes that permeate our literature and culture from Prometheus to Dr. Frankenstein to the golem to Dr. Faustus to John Hammond† to Dr. Venter. The hubris is the presumption that humans can create life, snatching unto themselves an ability that properly reposes only in the hand of God (or the gods). Following close on the heels of hubris, of course, comes nemesis, ready with those diamond chains to pin the new Prometheus to the rock of his never-ending torment. This myth is an enduring trope of science journalism, science fiction, and popular conceptions of science. That's the side of the story that sells newspapers and raises capital.

---

* More than $400 million, by some estimates.

† The fictitious billionaire who created Jurassic Park in the 1993 movie of the same name.

This modern myth has another dimension, though. The logic is familiar: if scientists can make life in the laboratory, this must prove that life could have originated from just the right chemistry, thereby proving that life needn't come from the hand of God. In our modern secular culture, this has sometimes emboldened the nonbeliever to smite the creationist, and with unseemly glee. The biblical injunction to be mindful of the plank in one's own eye (Matt. 7:3–5) is germane here, for Synthia presents an uncomfortable paradox for our atheist friends to contemplate. We might call it the "hand-of-the-scientist-god" paradox. Synthia carried a price tag of about $40 million. This money supported the numerous scientists, managers, and technicians involved, along with the highly sophisticated machinery and organizational infrastructure they needed to do their work. In short, Synthia was the collective product of the intelligence, foresight, and drive of everyone who worked to bring her into being. The paradox is that none of this even slightly undermines the creationist argument for the origin of life; it in fact strengthens it. Actual life did not need the JCVI to come into being, after all, nor any of its scientists, nor any of its sophisticated machines, nor any of its already-existing microbial helpers: it came about entirely on its own. How did *that* happen? If Craig Venter needed a platoon of the smartest people in the world to cobble together a poor imitation of life, just imagine the intelligent force that had to have brought the original into being!

———◆•◆———

What the hand-of-the-scientist-god paradox really tells us, of course, is that the origin of life remains an unsolved mystery. Perhaps it is unsolved because it is unsolvable—it is certainly not solvable by direct evidence or observation. No one was there to see it, and the Earth's tectonic churning has obliterated all the fossil evidence of its beginnings. There's a remote possibility we might catch its origin in action on other worlds, but we won't be visiting them in my lifetime, nor probably in the lifetimes of my children or my grandchildren. If clincher evidence is out there somewhere, it is likely to be a very long time coming.

This seems deeply unsatisfactory, so being the curious apes that we are, we peel back as many pages of life's history as we can, hoping to see what's printed on that first page.[2] This actually takes us quite close to where the origin must have occurred. We know from the likely age of the oldest microbial fossils that life was thriving on Earth by 3.4 billion years ago. We know that the Earth at that time was stumbling out from under an extended period of devastating meteorite bombardment.[3] It was hot and wet and there was no oxygen in the atmosphere. Dimly outlined on those early pages, like some ancient palimpsest, we can discern traces of some important milestones: cells practically from the beginning, the nucleus appearing about two billion years ago, embryogenesis and complex organisms about a billion years after that. But page 1 of the Book of Life, the page we're really interested in, remains plastered tightly against the thick front cover. It is unlikely that first page ever will be peeled away so we can read that first sentence, that first letter, or even, if you're inclined that way, to get a glimpse of whatever thought gave rise to that first letter.

The unknowable mystery of life's origins strengthens the various creation stories that are prominent features of the many religious faiths that bind and divide our species. We scientists don't like to put it this way, but we, too, have our creation stories—two of them in fact, both riding on the same level of intractable mystery.[4] On the one hand is the belief that life began with the origin of the gene: *genism,* let us call it.[*] The other believes that life's origin rests with the origin of the specified order-generating chemistry we call metabolism. There is no clever word to describe this way of thinking, so

---

[*] Robert Shapiro, in his provocative book *Origins,* refers to adherents to this faith as "naked genies."

we have to coin one: *physism,* from the same root, *physis*—nature, literally—from which we derive "physiology" and "physics," is as good a candidate as any.*

In the interests of full disclosure, I say at the outset that I incline toward physism—physiology was my intellectual mother's milk—but I am not really a strict adherent to that faith. Nor can anyone be, for that matter, for there is no room for fundamentalism, whether it be scientific or religious, on the question of life's origin. Life has two attributes that demand explanation, after all: the ability to store, transmit, and implement hereditary memory; and the sustaining, order-producing phenomenon that is homeostasis.

In current life, these two attributes are so tortuously intertwined that it is near impossible to imagine how one could have emerged spontaneously without the other being there first. Metabolism, to be more than mere chemistry, must be highly ordered, reliable, and reproducible. Bringing this orderliness reliably into being requires a high degree of specification, which must somehow be inherent in any presumptive living system. Currently, we think this specification inheres in replicators, in specific sequence codes of nucleotides in DNA that specify sequences of amino acids in proteins. So far, so good, but when we ask from where do the replicators themselves come, things begin to loop around on themselves. The replicability that underlies DNA's status as a repository of hereditary memory depends upon a host of metabolic processes specified by particular protein catalysts. Those protein catalysts would not exist, of course, without the

---

* I toyed for a time with the neologism *metaballeinism,* which derives from the Greek origin for "metabolism"—*metaballein: meta-*(over), *ballein* (to throw). From this, the word "metabolism" is rendered roughly as "changing," referring of course to the tendency of living systems to change one thing, say food, into another, say organism. The word has a subtle irony that I like: given the near theological divide between gene-first and metabolism-first doctrines, it is amusing to note that "metabolism" has a common etymological root with Satanism, with the word "devil" being rendered from *dia-*(across) and *ballein* to produce the root word *diabolo,* from which derives our modern word for devil as well as descriptive words like "diabolic." In the end, though, metaballeinism was just too awkward a word to love.

replicable hereditary memory. All this makes heredity itself a form of metabolism, and metabolism, in an odd way, a form of heredity.

The dilemma is obvious: each of the two necessary attributes of current life—heredity and metabolism—must exist for the other to exist. It is impossible (deluded, actually) to imagine such an intertwined system coming together all at once, with no intelligence guiding it. Yet if we are to believe that original life was anything like current life, we must believe they somehow did precisely that. To use a loaded phrase, present life seems to be "irreducibly complex."[*]

───────── • • ─────────

Is it possible even to have a coherent explanation for the origin of life with that dilemma sitting untangled at the heart of it? It's doubtful, I think, but that does not mean there have not been many ingenious attempts to untangle it.[5] Throughout the twentieth century, both physist and genist theories tied their fortunes to a still more primeval creation story, the doctrine of *spontogenesis,* which asserted that in the history of the Earth there had to have been at least one incidence of the spontaneous generation of life.[†6] It is a matter of scientific faith that this event could not have come from any overarching intelligence bringing it about: on this point, at least, you cannot simultaneously be a scientist and a creationist. Another tenet of the spontogenist catechism is that life began with a specialized sort of molecular churning from which, given just the right mix of lucky circumstances, life foamed into existence like an artfully whipped

---

[*] The phrase has come to be fraught because it is the very phrase advocates of intelligent design theory use to supposedly refute Darwinism. The phrase is nevertheless apt for the dilemma that is at the heart of the origin of life.

[†] Another school, pangenesis, argues that Earth was seeded with life by microbes transported through space on comets and meteors. It's certainly a credible idea for why the Earth has life, as we shall see, but it evades the question of the origin of life per se. See F. Hoyle & N. Wickramasing, "The Case for Life as a Cosmic Phenomenon," *Nature* 332, 509–511 (1986).

meringue. This prebiotic chemistry, as it's called, has provided both genist and physist grounds upon which many ingenious theories for the spontaneous origin of life have been built. These have brought us ever so close to the origin, but never quite to the starting point, at least not without having to verge uncomfortably into miraculous thinking—the same reliance on miracles that sustains creationists.

Nearly all spontogenist schemes for life's origin begin with the origin of the complex molecules that are the foundation of current life. By this, we mean four principal classes of carbon-based molecules: sugars (which provide fuel and structure), amino acids (which string together into protein catalysts), nucleotides (which string together to make genes), and lipids (which make membranes to envelop it all). Bring these together in just the right way, and you have life's essential infrastructure. A good chemist can, of course, whip up these components in the laboratory. But when life actually emerged, these components had to have arisen without the chemist's skill.

How they supposedly arose was through a long period of prebiotic synthesis that extended from the time the Earth cooled sufficiently to accumulate liquid water until life finally emerged. During this period, conditions on Earth were much different from the way they are presently. Then, the atmosphere was rich in the simple precursors of organic molecules, like carbon dioxide, ammonia, and methane. Add a little energy to the mix, in the form of lightning, or intense ultraviolet radiation streaming in from the young sun, and this could power the synthesis of molecules like cyanide and formaldehyde, which with a little bit more of a boost, could become the feedstock for amino acids, sugars, nucleotides, and a raft of other interesting stuff. If this went on long enough, the interesting stuff might have accumulated in a prebiotic "primordial soup."*

From there, the path to life is easy! All you have to do is string

---

* Another coinage of our friend J. B. S. Haldane.

these primordial molecules together in interesting ways: amino acids into proteins, which might give you a primordial metabolism; and nucleotides into nucleic acids, which might give you primordial genes. Then wrap them up in a lipid membrane, feed them some sugar—and off you go! I kid, of course, but this scenario, known as the Oparin-Haldane hypothesis,* has been the prevailing spontogenist paradigm for both genists and physists since the 1930s.

For many years, this scenario was another of J. B. S. Haldane's back-of-the-envelope ideas,† but in the 1950s, Stanley Miller and Harold Urey tried to cook up primordial soup in the laboratory, and they succeeded brilliantly. Miller and Urey built a mini-primordial Earth in a flask, with a tiny "ocean," "atmosphere," and "lightning." Left to run for a week or so, it produced all manner of interesting organic stuff. Since then, the Miller-Urey experiment has been the template for a veritable universe of variations, with the result that we have learned an awful lot about what prebiotic chemistry must have been like.‡

Unfortunately, among the awful lot of things we have learned is how flimsy a foundation prebiotic chemistry is for building a theory of life's origins. Worse, the more we have learned, the more daunting the problems have become.§ First, there is the problem of yield.

---

* Named after the Soviet chemist Alexander Oparin and J. B. S. Haldane, who conceived of it independently of one another. Haldane regarded it as almost a tossed-off idea and always graciously granted priority for it to Oparin, for whom it was a serious subject of lifelong study.

† What I would give to spend an evening in a pub with J. B. S. Haldane!

‡ Honors and accolades came to Miller and Urey in abundance, but the ultimate prize for this work eluded them. Miller and Urey were nominated several times for the Nobel Prize, but it was never awarded to them. Urey was a 1934 Nobel laureate in chemistry, however, for the discovery of the hydrogen isotope deuterium.

§ I can only scratch the surface here. There is a vast literature on this—too vast to summarize here, but it is summed up admirably and very readably in three books: Robert Shapiro's *Origins: A Skeptics Guide to the Creation of Life on Earth* (1986), Simon Conway Morris's *Life's Solution: Inevitable Humans in a Lonely Universe* (2004), and Iris Fry's *The Emergence of Life on Earth: A Historical and Scientific Overview* (2000).

What cooks up in a Miller-Urey flask is a diverse stew of organic molecules, most of which are uninteresting, with the interesting bits invariably present in quite small quantities. Imagine wanting to find a particular type of screw in a warehouse of brads, nails, nuts, clips, and a zillion other fasteners, all jumbled together in a mountain of little scraps of randomly shaped metal. How do you argue that a screw is in there at all—or in the warehouse of the primordial soup, how something lifelike can come from that? The answer is, "Not very plausibly."

There is also a process problem. Even a simple molecule like an amino acid is not synthesized all at once but is the end product of a series of reactions that have to take place in a particular sequence, each of which requires certain environmental conditions to come about. When you have enormous numbers of the precursors jumbling around, there is a finite chance that just the right sequence of events will occur; but the more interesting the desired product is, the smaller the chances become of it arising spontaneously. That is why the yields of the interesting things in a Miller-Urey experiment are so sparse.

Nucleotide synthesis provides a useful example: high temperatures are required for their abiotic synthesis, but the synthesized nucleotides will persist only if they are kept cold. The awkward question asks itself: how could both have come about on the hot young Earth? It's conceivable that the requisite high temperatures for synthesis could occur near, say, a hot spring, and that this could be adjacent to freezing cold temperatures—hot springs in Yellowstone Park in winter or in Iceland, to give two obvious examples. It becomes less plausible to imagine this for the prebiotic Earth, which was substantially hotter than the Earth is presently.

Finally (and I have by no means exhausted the list of objections), there is the "wild life problem." Prebiotic chemistry on the bench top enjoys numerous advantages that prebiotic chemistry "in the wild" would not have had: specified conditions, pure reagents in high concentrations, suppression of chemical "predators" that will react with and destroy life's fragile chemical precursors while still in the crib, a skilled chemist overseeing it all (unless we want to start thinking like creationists, that is). What would be the fate of incipient life in the wild, where no such protections could prevail? It would not be hopeful.

The default response to such problems has been to invoke the magic power of time: given enough of it, even highly improbable events, like the spontaneous and simultaneous origin of an enzyme and gene, might be expected to happen at least once. And only once, after all, is all we need. There's only cold comfort to be found in that idea, though, and it's becoming colder as those pages of the Book of Life have been peeled back closer to page 1. One can actually work out the probabilities for particular chemical events, say, a group of amino acids coming together to form a primitive functional protein. One can quibble over the computational details—scientific creationists tend to overstate the likelihood, although they hardly need to—but no matter how impeccable and dispassionate the calculation, they all lead to the same difficult conclusion. There is simply no way that the proverbial "lucky shot" would have happened, even over times that exceed the known age of the universe.

Even worse, as fossil traces of early life are beginning to be teased out, they are pointing to a surprisingly rapid emergence of life from the prebiotic Earth. On the basis of known times when the Earth could have cooled sufficiently to support life, and the subtle evidence of Earth's incipient life, the transition appears now to have taken about ten million to twenty million years, by some estimates. There may even have been multiple origins of life—dim embers of life emerging from the muck, only to be wiped out by an asteroid impact or some other catastrophe and replaced by the next glowing coal fanned into life.[7] That makes the origin of life a very fast proposition—the blink of a geological eye, or even, dare I say it, a rather easy thing to bring about.

The conclusion is inescapable: something beyond mere chance seems to have drawn life into being, helping it up from the dead world. But what could that something be? Creationists are at the ready with their answer, of course, waving their irrefutable claim for what (or, more precisely, who) did the helping.* You can scoff at their answer all you want, but that's just deflection from the embarrassing question: what is *your* answer?

———— ·•·• ————

The challenge for both genism and physism has been to bring both heredity and metabolism to a point where they are embodied in a single living entity, and without invoking squishy ideas like a guiding intelligence. The genist idea is a "replicator-first" theory, which equates the origin of life to the origin of the first so-called replicator, that is, something that can make replicas of itself, which implies a form of hereditary memory. The physist idea is a "metabolism-first" idea, which equates the origin of life to the first self-sustaining order-producing chemistry. This means that both physism and genism approach the same goal—mutually supporting heredity and metabolism—but from different perspectives. Ironically, both do so by blurring the distinction between heredity and metabolism.

The genist idea draws its breath from defining "replicator" very broadly.[8] A gene is a kind of replicator, as is a dividing bacterium, as is a pregnant elephant. All these replicators presumably evolved

---

* I hasten to add that "irrefutable" does not mean "correct," just incapable of refutation. It is revelation, not logic, that undergirds religious faith.

from more primitive replicators, which are descended from still more primitive replicators, which trace their ancestry to some original replicator—let us call it the *Ur*-replicator. If we have a good idea of what the *Ur*-replicator was, the spontogenist chemistry that gave rise to it might be worked out backwards from there.

Two theories prevail for what these *Ur*-replicators might have been. One, "RNA World," proposes that an alternate form of nucleic acid, ribonucleic acid (RNA), was the first replicator.[9] The second proposes an entirely novel form of replicator based upon clay crystals. The theories are not competing, as we shall see, but are complementary.

We'll take a stroll through RNA World first.* In present life, RNA is a form of "scratch-pad" memory that transcribes a snippet of DNA nucleotide code. With the aid of tiny organelles known as *ribosomes,* this snippet of nucleotide sequence can be translated into an amino acid sequence—a protein—that can do useful things. In RNA World, the *Ur*-replicator was supposedly a self-replicating RNA molecule that could carry out the catalytic functions of enzymes—a *ribozyme,* in a word.† Combining hereditary memory and function, as ribozymes do, meets the requirements for Darwinian evolution to occur, and once in place, life was presumably off to the races.[10] As the prebiotic chemistry of RNA World became ever more complex (it could go no other way), DNA and proteins eventually took over the respective phenomena of replication and function that had previously been unified in ribozymes. This left RNA bridging the emerging gap between them, as it now does in its role as a scratch-pad memory. In

---

* Walter Gilbert is credited with coining the phrase in the 1980s, but the idea has an older provenance that goes back to the origins of molecular genetics. Thomas Cech has also been a leading light in developing the presumed chemical scenarios that gave rise to RNA World, including the discovery that RNA could act as a "ribozyme," an RNA-based catalyst that could serve to mediate chemical metabolism in a self-replicating molecule, a discovery for which he shared the 1989 Nobel Prize in Chemistry with Sidney Altman.

† A portmanteau of *ribo*-nucleic acid and en-*zyme*.

short, RNA World was taken over by DNA World, which is the world we now inhabit.*

As the *Ur*-replicator, the ribozyme has a great deal to recommend it. Ribonucleotides are easier to synthesize and are more robust than DNA nucleotides are, and they can more readily assemble on their own into strings of nucleic acids. These attributes give RNA a slight leg up as the more likely nucleic acid to have first appeared in the primordial soup. Even more intriguing, small RNA molecules can self-replicate and can even catalyze chemical reactions, some of which can power spontaneous self-replication.[11]

Numerous serpents are lurking about RNA Eden, though: I will cite one big problem as an example.[12] RNA's ability to self-replicate can come off only if the molecule is small. Once RNA molecules get to be large enough to start doing interesting genetic and catalytic things, they tend to fold up on themselves in ways that prevent their self-replication. In the hands of a skilled chemist, these problems and many others can be worked around to yield impressive and tantalizing results. But the complexity of the fixes required tends to be self-defeating for RNA World to be the presumed original life—that hand-of-the-scientist-god problem, again.

This suggests that RNA World may have evolved from a still more primitive *Ur*-replicator, which is where clay crystals come in.[13] Let's set the stage. In its most basic form, the Darwinian idea couples some form of hereditary memory to some metabolic capacity to power replication of the memory. But is there really any reason that those things must be embodied in carbon-based chemistry, as they are in present life? Take replication. Replication of the DNA molecule involves more than a dozen specific reactions. The DNA

---

* This is demonstrated admirably by RNA's current role as "working memory" for the archived information in DNA, as well as by a class of viruses known as RNA viruses. These carry genes around in the form of RNA. When an RNA virus infects a cell, it uses the host to "back-copy" the viral RNA into DNA, which then goes on to infect the cell.

must be unwound, snipped in several places, new nucleotides slotted into place, and the whole thing restitched and wound back up. Each of these steps in turn requires many subsidiary and specific reactions to provide the cogs and wheels of the replication machine. For replication to be successful, all the cogs and wheels have to turn in synchrony. In present life, this is ensured by the specified catalytic milieu of the cell, which is largely the province of protein catalysts.[14] These are themselves specified by genes, which lands us right back in our irreducibility trap. Yet, need we insist that a catalytic milieu be proteins specified by genes? Not really. The catalytic milieu could be something else—like clay crystals.

Clay crystals as first life will be easier to understand if we know something about how catalysts work. Catalysts are devices that bias particular chemical reactions over others. If you want a reaction to go a particular way even if the reaction itself "wants" to go another, that is, it is thermodynamically favored to go that way, you need a catalyst. Catalysts pull off this trick through surfaces that bind particular molecules and line them up in particular ways so that electrons can easily move from one chemical bond to another. This is all a chemical reaction is, after all—getting electrons to leave one chemical bond and take up residence in another. By lining up particular molecules in particular ways, catalysts help bias one reaction—one particular type of electron movement, even an improbable one— over another.

Proteins (and RNA molecules too) work as catalysts by folding themselves up to form little pockets where molecules can nestle and mingle their electrons with one another. All catalysts work this way in fact, but the important requirement for catalysis to occur is that there *be* a nestling site. These can be provided in many different ways. Proteins do it by the way they fold, forming those little pockets. In industrial catalysts, the nestling sites are not fragile proteins, but robust ceramic surfaces, usually "doped" with interesting metal atoms to goose electrons along. The ordinary catalytic converter that cleans up unburned fuels from automobile exhaust is a familiar example. The heart of the catalytic converter is a ceramic element doped with platinum or some other precious metal like rhodium or palladium. The engine's hot exhaust gas raises the ceramic element to high temperature. When unburned hydrocarbons and carbon monoxide from the exhaust nestle into molecule-sized surface pockets on the ceramic, the embedded metal atoms help shift electrons between the unburned hydrocarbons and carbon monoxide to convert them into carbon dioxide and water.

"Hard" catalysts such as this may be the most likely candidate for RNA World's Ur-replicator.[15] This idea, put forth by the Scottish biochemist Graham Cairns-Smith, asserts that the first "living" systems were clay crystal "organisms." This seems like a plot for cheesy science fiction,* but it's weirdly plausible. Clays, like ceramics, are formed from dissolved silicate minerals that precipitate into crystals. Clays are extraordinarily diverse in what kinds of crystals they can form. Silicate crystals are also "dopeable," that is, amenable to the inclusion of various types of impurities called *dopants*. These can alter the silicates' crystalline structure, including the surface structures that mediate catalytic function. They can also provide ready pathways for electrons to stream from one reactant to another. Dopants need not be metals, as in a catalytic converter; they can be almost anything, including simple organic molecules, like those that might have existed in the chemical mish-mash of the primordial soup.

These silicon-carbon hybrids could have mediated specific catalytic functions, allowing them to, say, sort out the interesting chemicals from the dreck that accumulates in a typical Miller-Urey experiment. There's a bonus: because crystals are self-replicating,

---

* I can only offer a slim overview here of this fascinating idea. For more detail, see Cairns-Smith's two books *Seven Clues to the Origin of Life* (1985) and his more technical book *Genetic Takeover* (1987).

this highly specified catalytic chemistry could plausibly replicate as well. And there's icing on the cake: clay crystals can "compete" for a "food," namely, the soluble silicate and organic dopants floating about in the seas and ponds of the primordial Earth. The "winner" in any such competition grows by adding dissolved silicate to its bulk at the expense of "loser" crystals, which dissolve into solubilized silicate and dopant.

All this describes a primitive form of natural selection, and when you have natural selection, you have the mechanism for spinning ever more complex associations and hybrid silica-carbon "life forms." According to Cairns-Smith, this ratcheting complexity supposedly continued until the carbon side of evolving silicon-carbon hybrid "organisms" reached a point where there was a "genetic takeover"— the carbon-based side essentially kicking away the silicate scaffold that had supported it up till then. RNA World may have been the first of these to float out on its own. It's a crazy idea, but as is sometimes said, it may be just crazy enough to be true.*

But there's another side to the story. To be alive, these whiz-bang clay organisms needed a reliable source of energy and materials. In short, it is not sufficient to imagine a clay replicator of sufficient complexity to generate RNA World. We have to grapple seriously with the physist question: how do you get a metabolism that is sufficiently complex and robust to pull all this off? And without cheating?[16]

---

* The idea of silicon organisms has a venerable scientific history, probably being proposed first in 1891 by the German physicist Julius Scheiner but taken up in the 1920s most prominently by British scientists, notably H. G. Wells and J. B. S. Haldane. Silicon organisms and hybrid silicon-carbon life have also been recurring themes in science fiction, most prominently in Stanley Weinbaum's 1934 short story "A Martian Odyssey" (published in the book *A Martian Odyssey and Others* in 1949). Silicon-based life also played a role in *Star Trek* ("Devil in the Dark," 1967) as the Horta discovered on the planet Janus IV, and as hybrid silicon-carbon monsters that arose from a time warp in *The X-Files* episodes "Dreamland" and "Dreamland II," both broadcast in 1998.

As was the case for replicator-first theories, there are two principal strains of the metabolism-first idea. The first, which we may call *cellulist*, pins the origin of life to the origin of the cell. The second, which we will term *autocatalytic*, looks to chemical reaction systems that are capable of bootstrapping themselves into something like self-sustaining metabolism. Again, neither is exclusive of the other; they are complementary. Although cells probably followed the origin of metabolism, the cellulist idea came first historically, so we'll start there.

The primordial strain of cellulist thought had roots in the primordial soup. Oparin and Haldane were well aware that life could not have "just emerged" from the primordial soup. To both, the bridge from soup to life involved encapsulating primordial soup into tiny colloidal droplets they called *coacervates*.[17] There's nothing magical about coacervates—they can be whipped up in an undergraduate biology lab*—but what makes them interesting is that they segregate an internal environment from the surroundings by some sort of boundary made up of lipids, carbohydrates, or proteins or some similar substance.[18] Because coacervates are held together by strong ionic forces, they can pull electrons around and so catalyze selective chemical reactions. This is not a hypothetical ability: coacervates are widely used in the chemical, pharmaceutical, and food industries—as additives in industrial processes that can goose along chemical reactions that might otherwise be sluggish and even as "smart" drug delivery systems, in which they bind to surfaces of specific cells or tissues where the drug needs to be delivered.[19] So it is plausible to suppose that the selective chemistry wrought by coacervates could have helped life to foam up from the primordial soup.

The theory of coacervates as proto-cellular life probably reached

---

* See, for instance, "Creating Coacervates," at http://www.indiana.edu/~ensiweb/lessons/coacerv.html.

its peak sophistication in the 1970s, taken there by the chemist Sidney Fox as what he called his *proteinoid microsphere theory*.[20] Proteinoids are proteinlike polymers that string amino acids together through a novel mechanism. In the laboratory, proteinoids form spontaneously in a soup of amino acids mixed with phosphoric acid and elevated to high temperature, a tad less than 200°C typically. However, they can be made to form at temperatures as low as 70°C, which are well within the range one could expect to have prevailed on the early Earth (and indeed on present Earth, such as near hydrothermal vents or in hot springs). Proteinoids are not formed with the specificity of DNA-encoded proteins, of course, so they are a mixed lot. Nevertheless, they have the potential to be catalysts, although catalysts of what is a bit hit-or-miss.[21]

Proteinoids can associate with one another into coacervate-like globules called *microspheres,* and these have a number of interesting properties that bear tantalizing similarities to living cells. They are about the same size as bacteria, for example, and like bacteria, they sequester within them a sheltered environment that can catalyze chemical reactions, similar to what modern cells do.[*][22] Most interesting in this regard is that the inside and outside of the microsphere present *different* catalytic environments, raising the intriguing idea of coupled reactions, one outside and the other within, that could process materials through them, conjuring up visions of a primitive metabolism—and a primitive adaptive interface (Figure 10.1).[†] This proto-metabolism could include replication: proteinoid microspheres grow to a maximum size and then split in two. All dropletlike structures do this to an extent, but in the context of proteinoid microspheres as proto–life forms, it is tempting to look at this as a form of primitive cellular fission. The opposite is also true: microspheres

---

* Cell membranes are formed as sheets of lipids with both hydrophilic and hydrophobic domains.

† Proteinoids have the interesting capability of catalyzing the production of flavin, which helps manage energy metabolism in modern cells.

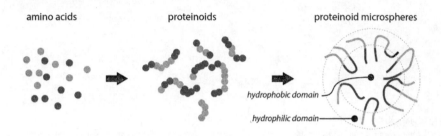

amino acids   proteinoids   proteinoid microspheres

hydrophobic domain

hydrophilic domain

Figure 10.1

Formation of proteinoids and proteinoid microspheres. Amino acids differ in their affinities for water: hydrophobic amino acids are dark gray while hydrophilic amino acids are light gray. These assemble into hydrophilic and hydrophobic domains that organize into microspheres.

can join and exchange material, dangling the even more tantalizing prospect of microsphere sex.

But, I digress. The proteinoid microsphere idea, indeed the whole coacervate idea, may be a plausible scenario for the emergence of a primitive cellularized metabolism,[23] but it, too, doesn't quite get us to original life. Some of the hurdles that must be cleared are easily dealt with, but they get increasingly difficult to clear the closer to the origin we go. A low hurdle, for example, is how to get catalytically diverse microspheres to mimic what present-day cells do, namely, to catalyze coherent reaction *systems,* that is, groups of particular catalysts that can reliably and persistently direct a particular and replicable flow of matter and energy through them.

An amino acid, for example, is not synthesized in one go but in several steps that must string together in a particular order. This means there must be a collection of catalysts that can manage the task as an assembly line does: one enzyme producing a product that is handed on to the next enzyme, which hands its product on to the next, and so forth until an amino acid pops out as the final product. Cells manage this effectively because the catalysts' properties are specified by genes, so that reliable sets of enzymes—the assembly lines—can be made, all encapsulated within the sequestered environment of

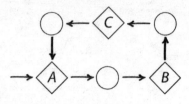

Figure 10.2

A hypothetical autocatalytic reaction system consisting of an array of catalysts (circles) and reactants and products for the catalysts (diamonds).

the cytoplasm. Coacervates, cleverly put together into "particles of life" called *jeewanu*,* can make simple amino acids, so it is conceivable that a collection of proteinoid microspheres with diverse catalytic capabilities could associate into assembly-line combinations of complementary catalytic systems, as bacteria do in microbial mat communities.

Without some form of heritable memory, though, it is very difficult to imagine any such proto-metabolic system ever stumbling over the threshold into life; as the small print often says, past results are no guarantee of future outcome. Without some form of embodied memory, the only thing proteinoid microspheres can ever do is churn the primordial soup. Churning, even sophisticated and complicated churning, is not life. Embodied memory is where the second strain of physist thought comes in: autocatalysis.

Autocatalysis is kind of chemical recursion that confers persistence to a reaction (Figure 10.2). Imagine a chemical reaction, *A* to *B*, in which *B* reacts to form an intermediate product, *C*, which feeds back onto *A*. This recursion makes the reaction sequence autocatalytic because the formation of *B* ultimately promotes the formation of the initial reactant *A*. We could imagine, for example, that *A* is an

---

* Jeewanu, Sanskrit for "particles of life," is an idea formulated in the 1950s by Krishna Bahadur of the University of Allahabad. It was long forgotten but recently championed by Mathias Grote of the University of Exeter in "Jeewanu, or the 'particles of life': The approach of Krishna Bahadur in 20th century origin of life research," *Journal of Bioscience* 36, 563–570 (2011).

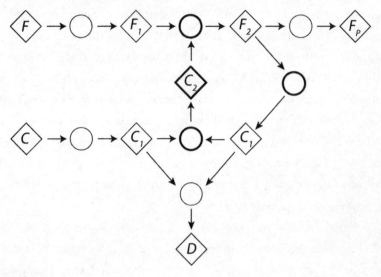

Figure 10.3

An autocatalytic loop as a form of proto-metabolism. Circles represent catalysts, and diamonds represent reactants and products. The autocatalytic loop is represented in bold.

amino acid floating about in the primordial soup, and B is a peptide that grows by adding amino acids to it. If the peptide B produces the precursor to an amino acid C, this will form more amino acids, which promotes the growth of the peptide B. In short, the reaction catalyzes itself.

What is significant for our discussion is the fact that autocatalytic reactions are *persistent* processes, and this makes them a form of memory capable of biasing the future to one outcome out of many possible alternative outcomes. Imagine a simple reaction system where reactants C and F feed into a system of catalytic surfaces, with three possible products, C_2, D, and F_p (Figure 10.3). If there is a suite of catalytic surfaces that favors a closed loop of reactions, there will be a bias to the production of C_2, disfavoring the other outcomes. Thus, the autocatalytic loop will capture a stream of C and F to boost the formation of C_2. This is a persistent form of metabolism.[24]

Like coacervates, this is not a far-fetched idea: bacteria synthesize various types of protein antibiotics through just such a process.[25] If such processes are scalable, that is, they can be made to operate at larger scales than the molecular, then you have a conceivable pathway to, say, RNA World, or to jeewanu, or to autocatalytic sets of complementary proteinoid microspheres.[26] Whether or not the pathway is feasible turns crucially on how sustainable the process is. To become metabolism, these pathways must work together to support homeostasis of some sort. To become life, they must simultaneously embody hereditary memory. The hereditary memory of any reaction system is its persistence.

The real value of the autocatalytic idea is that it dissolves the division between the genist and physist camps. There is now no longer any distinction to be drawn between metabolism-first and gene-first theories because metabolism and hereditary memory are one and the same. This logic has led an Israeli chemist, Addy Pross, to suggest that life emerged as a kind of proto-homeostasis that he calls *dynamic kinetic stability,* or DKS. In a nutshell, Pross's DKS refers to a balance between order-forming sets of chemical reactions and their inevitable degradation. According to Pross, chemistry became biology through a ratchet of ever-increasing complexity of systems of dynamic kinetic stability. Eventually, like Cairns-Smith's clay organisms, the ratchet eventually lifted chemistry into life.

We are familiar with this idea: it is the same idea as the persistent dynamic disequilibrium outlined in Chapter 2.* Where Pross's idea stands out is his thermodynamic conception of life, metabolism, and hereditary memory. There is a wrinkle to keep in mind, however. DKS is basically a theory of a "cumulus cloud" type of order-producing thermodynamic system, as outlined in the Preface. The question posed in the Preface stands: how does such a system convert

---

* I have a small quibble with describing such a system as dynamic kinetic stability, as a disequilibrium is anything but stable.

itself into a "cauliflower"-type of living, order-producing thermody-namic system? That remains unsolved.

So, after several decades of ingenious effort (which I've only very superficially summarized), where do we stand on the question we started with: how did life emerge from nonlife? Let me put the question another way, similar to the one I posed in the Preface. How would we know life if we had been there to see it emerge? To resurrect the metaphor, how could we see "cauliflower" life emerge from the "cumulus cloud" imitation life? When could we say that life had emerged from what must have been a gray area where "cumulus cloud" merged imperceptibly into "cauliflower?" Without a coherent idea of what life is, the question cannot really be answered. We've gotten ever so close, but not all the way.

Perhaps, then, the time is ripe for thinking about it another way. My suggestion is that we are thinking about it on the wrong scale. Ever since Darwin's "warm little pond,"* ever since Thomas Huxley's and Ernst Haeckel's protoplasm,† spontogenist, genist, and physist have all looked for life to emerge from the bottom up. We start with the uttermost fine details of chemistry, energy transfer, and tiny

---

* In an 1871 letter to Joseph Hooker speculating on how life itself could have originated, Darwin wrote: "It is often said that all the conditions for the first production of a living organism are present, which could ever have been present. But if (and Oh! what a big if!) we could conceive in some warm little pond, with all sorts of ammonia and phosphoric salts, light, heat, electricity, etc., present, that a protein compound was chemically formed ready to undergo still more complex changes, at the present day such matter would be instantly devoured or absorbed, which would not have been the case before living creatures were formed."

† *Protoplasm* was believed to be a prebiotic ooze (the literal meaning of protoplasm) that was the quasi-living foundation of life. This was in keeping with Huxley's championing of doctrines of spontaneous generation, and Huxley even claimed to have discovered protoplasm in deep ocean sediments. Huxley's protoplasm turned out to have been a preservation artifact. For a fascinating history of doctrines of spontaneous generation and their relationship to Darwinism, see Strick's *Sparks of Life: Darwinism and the Victorian Debates over Spontaneous Generation* (Harvard University Press: 2002).

catalysts, and look to life to emerge from there. There's one very large problem with looking to the small scale for life to emerge, however, and that is the relentless scourge of the small scale: diffusion.

To us, large creatures accustomed to perceiving the world at a large scale, diffusion is a benign, weak force. It is diffusion that drives the gentle drift of perfume across a room, the languid spread of color through water from a blotch of dye. At the small scales where we imagine life to have emerged, however, diffusion is literally an explosive phenomenon, scattering everything before it into oblivion: incipient metabolic networks, incipient genes, incipient cells. At the small scale of actual cells, this explosive force is shielded by structure, by enclosing the multitudinous and delicate molecules of the cell's catalytic milieu behind the bulwark of the membrane, which at this scale is extraordinarily robust. Before there were cells, though, Darwin's "warm little pond" was a violent, storm-tossed place for delicate molecular proto-life. How could proto-life ever have managed to survive in such an environment, let alone evolve into life? It's nearly impossible to imagine, and this is why *every* theory that imagines life to have emerged from the "bottom up"—from the molecular scale—has come up short.

This leads me to a strange thought. As I write these words, I am surrounded by a garden of grass, trees, flowers, and herbs that is visited from time to time by animals that eat the plants and (rarely) by other animals that eat the first animals in turn (I'm particularly mindful of a black cat that visits from his cosseted home next door to pick off the birds in my garden). In the soil live innumerable bacteria and fungi leading busy lives consuming what the animals have left. It is a hectic, tangled world of little lives.

As I allow my physiologist's mind to drift, a new picture emerges of what I see. What really surrounds me (and carries me along as well) is not quite a complex world of *things,* but a complex *cascade of energy:* a turbulent, frothy wave of electrons interposed between the sun and the rest of the universe. The wave crests first at the interception of light by green plants,* tumbles down the curl through animals, and ends with the bubbly foam of the decomposer bacteria and fungi.† If you surfed along with this wave, you would see the complex structure of the wave surface bubbling, foaming, circulating, reflecting the innumerable fractal surfaces that manage the energy flow: the folded chloroplasts, the spongy interiors of leaves, the vast surfaces of lungs and intestinal tracts, the structured soils—energy cascading through them all forward in time.

This foamy wave of energy is a standing wave, like the wake that follows a speedboat. It exists in time and persists as long as energy is being fed into it. The standing wave is a metaphor for the open thermodynamic system I introduced in the Preface: energy flows in, does work, and then is dissipated as heat. On Earth, the standing wave has existed as long as the sun has spewed forth photons and the Earth has been present to intercept them. The standing wave exists in some form wherever any planet is close enough for the sun's energy to do significant work on it.

Energy flowing through open thermodynamic systems often imposes orderliness on them.[27] There is nothing magical about this; the orderliness is simply the fastest way to hurry energy through the standing wave, from photons from the sun to heat that warms the universe.[28] We can see this orderliness in the stately waves that roll across oceans and atmospheres, manifest as great circulation cells, by large breakaway eddies that boil occasionally here and there into chaotic storms. Such patterns are also evident on Mars and Venus, in the tumultuous weather of those planets.

---

* At a global rate of roughly 90 terawatts, or about 0.07 percent of the total energy from the sun.

† Boris Pasternak captures this in an evocative passage from *Doctor Zhivago:* "Everything around fermented, grew, and rose on the magic yeast of being. The rapture of life, like a gentle wind, went in a broad wave, not noticing where, over the earth and the town, through walls and fences, through wood and flesh, seizing everything with trembling on its way."

What life on Earth has wrought is a change in the wave's form. You can see the difference if you look at it the right way. A Martian landscape and a barren landscape on Earth look disturbingly similar, shaped by physical forces of erosion and soil movement. Wind and water erosion patterns on Earth and Mars resemble one another closely, for example. So great are the similarities, in fact, that NASA uses several of these terrestrial landscapes—in the Namib Desert or the Atacama Desert, for example—as mock Martian landscapes to test rovers and other devices for deployment to Mars. Compare this to a life-rich landscape on Earth, and the difference is clear: the Earth landscape is much richer and filled with more complex shapes.[29] That leaf is not so much a thing as it is an interface between photons streaming in from the sun and ultimate disorder and heat—its complex form sustained by the leaf's ability to capture energy and funnel matter through it and shape its progress to disorder in a highly specified way.

Although the form of orderliness might differ, the important point is that there is orderliness in the first place. Let us reflect on why. We spoke in the preface of the spontaneous orderliness of open thermodynamic systems, the Fourth Law of Thermodynamics. Orderliness in open thermodynamic systems is a form of work: it takes energy to sustain it. At the same time, orderly systems spontaneously decay to disorder, usually wrought by the disruptive power of diffusion. That is the Second Law of Thermodynamics. Orderliness in an open thermodynamic system therefore persists only when sufficient order-generating energy flows through to overcome diffusion's disruptive power. This is why the peculiar orderliness of weather and cumulus clouds is a phenomenon of large scale. At the scale of the cell, there is no such thing as weather, only destruction.

Let's picture this in our metaphor of a standing wave of energy flowing through the open thermodynamic system of the Earth's (Figure 10.4).[30] The Earth intercepts a continuous stream of energy

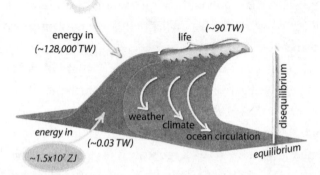

Figure 10.4
The standing wave metaphor for the Earth's open thermodynamic system. Details in text. Figure taken from the Stanford Global Climate and Energy Project.

from the sun, at a rate of about 128 petawatts.* By way of comparison, the world's fossil fuel energy consumption amounts to around 16 terawatts, roughly 0.0125 percent of the energy streaming in from the sun. The energy streaming through all life on Earth amounts to about 90 terawatts, which is also about 0.01 percent of the light energy that actually reaches the surface of the Earth, fueling plant growth. All the rest of the energy streaming in to the Earth—the other 99.99 percent—powers the orderly patterns of global atmospheric and oceanic circulation that drives global weather.

Life, therefore, represents a tiny froth on the crest of the Earth's standing wave of orderliness. The point is that much of the orderliness of life comes from orderliness already existing on the open thermodynamic system of the Earth.[31] The signature of life amounts to a specified orderliness that shapes the crest. What life has done

---

* A petawatt is a very, very large flow rate of energy, 10^{15} joules of energy per second. That's a "1" followed by fifteen zeros: 1,000,000,000,000,000 watts per second. Very large numbers like this are usually prefaced with a prefix ordered by multiples of 1,000. A terawatt, for example, is a flow of 10^{12} joules of energy per second. A petawatt is equal to 1,000 terawatts. An energy influx of 128 petawatts from the sun is equal to 128,000 terawatts.

is to make the wave crest "foamier," if you will, ordering it into an infinitude of nested and interlocking systems of persistent and specified energy flows, vast conspiracies of life that currently envelop our living planet.[32]

Which leads us to the strange question: what law demands that life has to evolve *up*, from the small scale to the large? Why couldn't it have been the other way? Why couldn't life—homeostasis, essentially—have emerged first at the large scale, even as a *planetary* phenomenon, sustained at a large scale on pre-existing orderly flows of matter and energy until it could be encapsulated within the safe harbor of the cell? All that is needed is an energy source that is large enough to overcome the disruptive power of diffusion at a small scale and that is persistent enough to allow incipient conspiracies of homeostasis to piggyback on that standing thermodynamic wave. And that only occurs at *large* scale.

The energy needn't come from the sun, which is what mostly drives present life. One could imagine any number of other large-scale energy sources that might do. Residual heat from the formation of the Earth, emerging, say, in concentrated form around hydrothermal vents might do.[33] Natural fission reactors, like the one discovered in Gabon that generated power at a rate of about 100 kilowatts for roughly three hundred thousand years, might also serve.[34] No matter what the putative energy source, this thermodynamic approach brings a distinctive perspective to the problem of the origin of life: it turns Darwin's "warm little pond" upside down, because it is only at a large scale that life—that is, homeostasis—can arise on its own, without help from the hand-of-the-scientist-god.

The idea of life originating at a planetary scale is odd enough, but if we follow the logic that led us here, we can follow it further to an even stranger thought. Let's first recap where we have come. If Biology's Second Law is true and life is at root an expression of the phenomenon of homeostasis, then the origin of life is tantamount to the origin of homeostasis. Homeostasis demands certain things, however—among them at least rudimentary forms of cognition and intentionality. This leads to the very strange thought that the origin of life is tantamount to the origin of cognition and intentionality. Even stranger, cognition and intentionality had to have actually *preceded* the origin of cellular life.

We can rescue this idea from the loony bin by defining cognition very broadly and generally—as informing a state or process about its environment. Our own very complex cognition should not blind us to the fact that cognition can be framed even in very simple systems, like individual cells, or even simpler. The nerve cells that underlie our own cognitive systems are certainly cognitively aware, but they are cognitively aware in a very different and highly circumscribed way from the large-scale cognitive phenomena in which they participate. An individual nerve cell is cognitively aware of the fluid environment in the brain in which it bathes, and of the chemical signals flung at it by the myriad other nerve cells communicating their own cognitive states, and very likely many other features of its little world. Outside of brains, individual cells are cognitive entities in the same way. A photosynthetic algal cell maps the presence or absence of light onto its encapsulated catalytic milieu, altering the cell's physiology in accordance with its environment.[35]

Similarly, intentionality can be defined very broadly. As I argued in *The Tinkerer's Accomplice,* intentionality can be construed as the coupling of cognition to metabolic engines that can shape the world to conform to a cognitive map.* Brains produce a very complicated intentionality. If I have a cognitive vision of my office being organized

---

* This is the argument that got me in trouble with that disgruntled reviewer mentioned in the Preface.

in a particular way, I can do the work to bring my office into conformity with that cognitive map. When my office degrades into inevitable chaos, I do the work again to conform it to that cognitive vision of an orderly office. There is no reason to suppose that this kind of intentionality cannot operate at different scales of life. The microbial mat, for example, is the large-scale emergence of a constructed environment that reflects the awareness of each species of microbe of its local environment, and the reshaping of that environment to bring it into conformity with the microbe's internal cognitive map of what its surroundings should be.

If we imagine what those first conspiracies of homeostasis must have been like, then we can envision a kind of selection operating between different persistent systems of energy and mass flow. Fitness in this instance will be equivalent to persistence, and persistence will be equivalent to robustness of homeostasis. Persistence will turn on how effectively the present and persistent state of an order-producing thermodynamic system—a kind of memory, keep in mind—can shape its broader environment to bring it into conformity with itself. Fundamentally, this is coupling work to information, which is what relates intentionality and cognition in purposeful living systems. This implies that cognition and intentionality were with life from the get-go, and before cellular life emerged from the living thermodynamic froth.

The origin of life invites us to think differently about the nature of life and its distinction from the material world. Specifically, thinking of life as chemistry or simple mechanism has led us to ingenious insights into the origins of self-replicating chemistry and the fuzzy boundary between the twin pillars of the Darwinian idea: apt metabolism made reproducible by heritable memory. It has led us ultimately to a dead end, however, because life emerging from the small scale to the large runs up against the formidable hurdle of diffusion. Confronted with this hurdle, we are forced to turn the problem upside down. In so doing, however, we are drawn to think about life as a global phenomenon, driven by the emergence of homeostasis and the cognitive capabilities that implies.

Among the different perspectives is that life shapes its world according to its desires and striving toward homeostasis. Life, in other words, emerged as a massive extended—an intentional—organism.

11

PLATO STREET

On the southern outskirts of Windhoek, the capital city of Namibia, there sits a modest little neighborhood called Academia (Figure 11.1). It is so named because of its proximity to the campus of the University of Namibia, chartered in 1993, three years after Namibia's independence. Academia's claim to distinction is naming all its streets after famous philosophers, hoping, I suppose, to inspire the new scholars that would swarm to the new university with thoughts of Great Things.

The layout of the streets offers some unintentional philosophical humor. For example, the neighborhood's trunk road is Plato Street. This is sensible, I suppose, in light of Alfred North Whitehead's famous aphorism that all of philosophy is a series of footnotes to Plato. Yet Socrates, arguably more the root of Western philosophy than his prolific scion Plato, gets allocated only a minor side street. Aristotle Street, meanwhile, is a shabby peripheral avenue that faces onto Windhoek's noisy municipal airport. Is this a subtle commentary on the conflict between Plato and Aristotle, which Aristotle lost and which led to him decamping in sulky self-exile to Asia Minor? More germane to me, the biologist, is the short shrift given to Charles

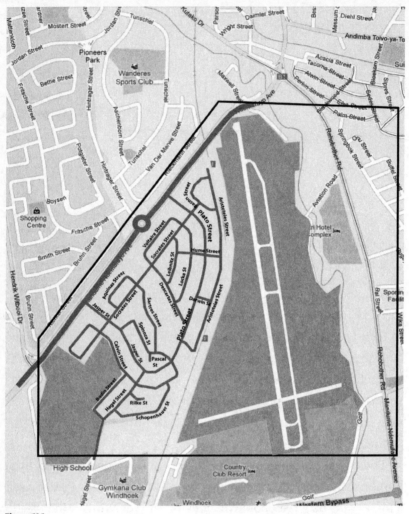

Figure 11.1
Academia.

Darwin. If I had been the neighborhood's developer, I would have named the major thoroughfare after him, not Plato. Instead, Darwin Street is an insignificant connector between Plato and Aristotle Streets, just long enough for two homes on each side of the street (Figure 11.2). One wonders whether the developers gave any thought at all to the layout of their presumptuous little neighborhood.

Figure 11.2
Plato Street, in the Academia neighborhood of Windhoek. *(a)* Intersection of Plato Street and Darwin Street. *(b)* The view of Aristotle Street. *(c)* Darwin Street, looking toward Aristotle Street from Plato Street. The Windhoek Municipal Airport is visible at the end.

Of course they didn't, and it's silly even to ask it. But it gives me an excuse to ask an interesting question: where, philosophically, does biology sit? Does it sit with Plato and his notion that the universe is motivated by striving toward his ethereal universe of ideals? That's where biology sat for much of its history, from Plato himself through Linnaeus and his Platonic theory of the species through William Paley. Or would biology sit more comfortably with Aristotle? That seems more likely, as a clear thread still connects Aristotle's notions of internal strivings with medicine, physiology, and embryology. Or perhaps biology sits with neither of these purposeful avatars, as the modern legatees of Darwin's name would assert? If that is true, then what is Darwin Street doing bridging Plato and Aristotle Streets? Wasn't Darwin supposed to have broken us free from all this purposefulness stuff?

The point of my little soliloquy on streets and philosophers is that eventually life, mute and insistent, inevitably confronts us with the perplexing riddle it has always posed. To mangle a common joke, Why did the chicken cross Plato Street? The scientist's reflex is immediately to recast the question—*How* did the chicken cross Plato Street?—in the hope that the rephrased question will give us a "real" answer: because its legs carried it there. Modern science has been drawn into that unfunny answer because it opens the gates to the lush landscapes of the "how": what controls the legs' movements, how the muscles attach, what the circuit diagram looks like for the chicken's walking. Yet sitting just outside those luxurious depths lurks what is arguably the only "real" answer to the original question: because it *wanted* to.

———— • ————

Wanting something is desire, and meeting wants is purpose, and both are inextricably bound up with "why." But is it possible— *permissible,* even?—for a scientist (a "real" scientist) to think about purpose and desire in any other way but "how"? Certainly, for scientists, the "how" of purpose and desire has been the more scientifically fruitful question than the "why." We have begun to map out the neural architecture of emotions, motivations, and intentions, at least for the brains of humans. We know the parts of the brain and the neurotransmitters involved; we can manipulate emotions and desires with pharmaceuticals. We can evoke remarkably vivid sensations and wants by tickling just the right parts of the brain with electrical currents. We can read the weak magnetic fields that emanate from the meat-ware computers crackling away inside our heads. We can even begin to discern the thoughts that are generating them. But has our increasingly sophisticated ability to map the three-dimensional architecture of desire brought us any closer to the "why" of desire?—which would seem to be the most important thing about it. As some wag once put it, a computer called Watson may have beaten Ken Jennings at *Jeopardy*,* but did Watson *want* to win? Did Watson *know* he (it) won? Even granting the dubious long shot that Watson could be said to *want* anything at all, Ken Jennings undoubtedly brought desire to the game in an entirely different way from Watson. So the question remains: even if we understood in the uttermost detail *how* the chicken crossed Plato Street, would that entitle us to say anything—anything at all—about *why* the chicken crossed Plato Street?

The problem of purpose and desire is thorny enough for life as it exists, but it becomes an enormously more difficult proposition when it comes to evolution. Individuals may have purpose and desire. We know this of ourselves, undeniably, and we can guess pretty well that we see purpose and desire in fellow humans. But it is a tenet of received Darwinian wisdom that individuals die, leaving only their lineages to evolve. What role could desire possibly play there, when desires must die with the individual feeling them? Or do their desires really die? Remember, this was the sticking point between Lamarck and Darwin. Lamarck thought that the ineffable forces that drove purpose and desire in individuals *could* transmit through lineages. Darwin didn't think much of Lamarck's ineffable forces, but if his Lamarckian schemes of inheritance bear any witness, he was attuned to adaptations being transmitted along lineages. Although we might quibble that Darwin's disagreement with Lamarck was not quite as deep as we imagine today that it was, it is incontrovertible that, with

---

* Watson was a computer system built by IBM, and Ken Jennings won more than $2.5 million on *Jeopardy*.

the advent of Neo-Darwinism, the break became total. Ever since, the a priori exclusion of the "why" has become so forceful that even bringing it up elicits something akin to religious warfare.[*]

Adaptation, the problem with which we began this book, is the essential "why" phenomenon of evolution. The modern synthesis claims to have an answer to the problem of adaptation, but its solution is framed in terms of "how": adaptation comes about from the selection of apt function genes. Yet without the "why," the "how" solution is an empty tautology.

This conundrum was familiar to the great founders of the modern synthesis, and they grappled to find a way out of it. Most notable among them was Sewall Wright, to whom we have already been introduced. His solution to the conundrum of adaptation bequeathed to us one of the most powerful metaphors of modern Darwinism, the *adaptive landscape*. The adaptive landscape metaphor, in turn, bled over into ecology as a related metaphor, the *niche space,* which laid out a powerful conceptual model of adaptation to the environment. Taken together, these two metaphors have provided ecologists and evolutionists a common and fruitful way to think about

---

[*] Exhibit A on this score is the near-hysteria that recently gripped evolutionists worldwide over intelligent design theory (IDT). Looked at objectively, IDT is a rather harmless and benign resurgence of Neoplatonism. Yet it was commonly represented in the scientific "community" as something akin to the Golden Horde storming the Gates of Vienna, with all the illiberal responses one expects in a community that perceives itself as under siege. The controversy was more or less suppressed with a federal judge's 2005 ruling in *Kitzmiller v. Dover* that IDT is not science and therefore was proscribed from being taught in science classrooms. The irony of "academic freedom" seeking protection behind a federal judge defining what science is was lost in the victorious celebrations that followed the ruling. The story of *Kitzmiller v. Dover* is admirably laid out in Edward Humes's *Monkey Girl: Evolution, Education, Religion, and the Battle for America's Soul* (2007), with another take by yours truly in *The Christian Century* ("Signs of Design," Vol. 124, 18–22 [2007]). *Kitzmiller v. Dover* did not sate the militant secularism that drove it, however. This is evident in a more recent bit of secularist hysteria from Arizona State University physics professor Lawrence Krauss, who opines in the September 8, 2015, issue of *The New Yorker* that—well, the title says it all: "All Scientists Should Be *Militant* Atheists" (emphasis added by me).

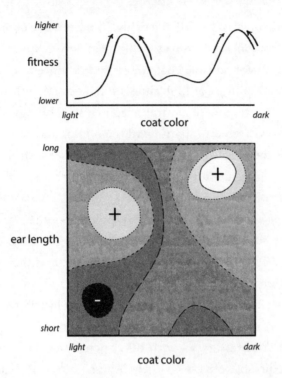

Figure 11.3
Fitness landscapes. *Top:* A one-dimensional fitness function. *Bottom:* A two-dimensional fitness function. Details in text.

the problem of adaptation, and in a way that purportedly vaults over Plato's purposeful ghost. I argue, however, that it does just the opposite: it obscures the problem of adaptation even more.

Wright's solution to the problem of adaptation was conceptually simple and elegant, and as you would expect, couched deeply in mathematical reasoning. We'll just stick as much as we can to the conceptual here, if you please, which Wright called a *fitness landscape*. Wright's fitness landscape begins, as most maps do, with a line. Imagine some trait—let us say coat color—that varies from light to dark. We represent this variation along an axis, from light to dark, and each coat color along this axis carries with it some consequence for the coat-bearer's fitness (Figure 11.3). Some coat colors—a dark coat in a dark forest, say—will mean high fitness, while other colors—say, a light coat in a dark forest—will mean lower fitness. Mapping fitness over all conceivable coat colors will produce a squiggly line of fitness, with humps of high fitness interspersed with troughs of low fitness. In a regime of natural selection, Wright argued, populations will gravitate naturally toward the humps and away from the troughs, for the simple reason that fitness is higher at the humps and lower at the troughs. In this way, those genes that specify the particular coat colors that are associated with high fitness will come to prevail at the expense of those genes that specify other colors.

Adaptive humps and troughs are a simple graphical representation of Wright's and Ronald Fisher's mathematical models for gene selection. Where Wright and Fisher differed was how multiple gene-specified traits could interact to affect fitness. Were genes selected more or less independently of other genes (Fisher), or were there interactions of genes with other genes that together would have unpredictable effects on their fitness (Wright)? This dispute simmered throughout their lives. But the differences between the two men are not so interesting to us as Wright's description of how to assess the fitness consequences of multiple interacting genes. His description became the metaphor known as the adaptive landscape.

The squiggly line of fitness I've just described is the simplest adaptive landscape, namely, the fitness consequences of variation of a single trait, plotted graphically along one axis (coat color), or to phrase it properly, one trait dimension. We can easily expand that to consider the fitness consequences of two traits together, let us say coat color and length of the ear (see Figure 11.3). All the possible combinations of coat color and ear length can be mapped out along two axes (properly, two dimensions), oriented perpendicularly to one another: dark to light coat color on one axis and short to long ears on the other. Instead of a squiggly line of fitness, we now have a *surface*, on which

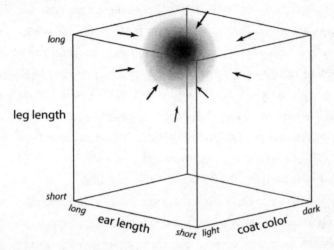

Figure 11.4
A three-dimensional adaptive space.

each combination of ear length and coat color will be represented by some point hovering above the plane. Together, these hovering points map out a fitness landscape. In the same way a topographical map represents the peaks and lowlands of a landscape, the adaptive landscape maps out adaptive highlands and maladaptive lowlands. Natural selection would then steer populations toward the adaptive peaks of survival and reproduction and away from the adaptive valleys of death.

Here is where the adaptive landscape metaphor begins to get really interesting, for mathematically, there is no reason to stop at two traits. What about three? We simply now plot three trait dimensions against one another, with the three traits plotted on axes that are all perpendicular to one another (Figure 11.4). Now, instead of a sheetlike fitness landscape as with two dimensions, there is now a fitness "space" that is enclosed within the cubic volume bounded by the three trait dimensions, with coat color varying along one horizontal axis (the length), ear length along another horizontal axis (the depth), and (as long as we're making this up) leg length along a vertical axis (the height). Again, it's easy to spin scenarios of the likely fitness consequences of the three traits together. Animals with dark coat color and long ears might be fitter if they had long legs compared with animals with short legs. It's a fitness trifecta: they blend in, can detect hungry predators sooner, and can run away faster. I could go on spinning adaptive tales, but you get the point. Now, gene combinations that produce high fitness appear as "adaptive spaces" rather than adaptive peaks. Under a regime of natural selection, populations will gravitate over time toward these local spaces of high fitness, just as moons might gather around a planet.

An adaptive space in three dimensions is easy to visualize, but there's no mathematical reason why we need stop at three. Mathematically, we could map four, ten, a thousand traits against one another, and the end would be the same. We could have four-dimensional, ten-dimensional, thousand-dimensional, indeed n-dimensional *adaptive hyperspaces* that could account for any number, n, of traits that vary with one another. The number of dimensions poses no problem, because mathematicians have clever tricks for dealing with such cases. Although it becomes progressively more difficult to mentally visualize an adaptive hyperspace as more and more dimensions are added, an adaptive n-dimensional space could still be calculated by a computer so that interesting things could be said about it. We could point, for example, to coat color, ear length, leg length, folding pattern of the brain, and density of the mitochondria in the muscles as five interacting traits. These would produce a five-dimensional adaptive hyperspace, and if we were clever enough, we could calculate a point, or a number of points, in this five-dimensional hyperspace toward which lineages would gravitate for maximum fitness.

It is a beautiful idea, guaranteed to send shivers down the spines of the mathematically inclined. Most beautiful is that it implies that only a limited set of gene combinations can evolve, compared with the infinitude of possible combinations. To the extent that these gene combinations correspond to particular types of organisms, the adaptive hyperspace limits the kinds of genetic organisms—the genetic combination of traits—that are possible.

———•———

The conceptual beauty of Wright's adaptive hyperspace metaphor has spread widely beyond evolutionary genetics, perhaps most significantly into ecology, where it has become transformed into the parallel conception of the ecological niche. The ecological niche is a venerable idea that predates the Neo-Darwinian synthesis.[1] In the early part of the twentieth century, ecologists had begun to conceptualize nature as comprising species occupying "niches," that is, pursuing ways of life that distinguish one creature from another and that describe their natural places, in either space, time, or habit. Different species of birds, for example, might occupy different tiers within a forest canopy. One species might glean insects from trunks of trees, another might flit about the darkness of the lower stories of the canopy, and still another might forage in the sunnier levels higher up. Larks and swifts might glean insects from the sky during the day, while bats might come out at night to do the same thing. You get the idea. As it was originally conceived, the niche concept was reminiscent of William Paley's well-ordered creation—every species in its place, together producing a harmoniously functioning ecosystem. Phrasing it as an ecological "niche" presumably allowed this obvious orderliness of nature to be expressed without any inconvenient Paleyesque natural theology.

Like nearly everything else biological in the early twentieth century, the Neo-Darwinian revolution dramatically transformed this idea of the well-ordered niche. No longer was the niche an expression of a creature's proper place in nature: it became a site of contention and competition, tied up in nature's tragic drama, red in tooth and claw. Those lineages lucky enough to occupy a niche would have to

Figure 11.5
George Evelyn Hutchinson (1903-1991), consulting a book in
his office. Hutchinson was thirty-six when this photograph
was taken; by that time, he had settled at Yale University.

defend it, excluding presumptive interlopers or pushing them away
to make them find their own niches, or coming to some uneasy
accommodation to divide the niche with them. As this transformation was unfolding, the niche concept went through several itera-
tions: the niche was habit, the niche was space, the niche was time.

The genius who finally made sense of it all was another of those
Cambridge polymaths, George Evelyn Hutchinson (1903–1991).[2]
Hutchinson was an ecologist, drawn into that field by an early interest in aquatic beetles and how they fit into the simple aquatic systems
they inhabited. Born in Cambridge, England, to an academic family,
his childhood playground was, in his words, "inhabited by genius."
His early childhood education at a prestigious British prep school,
Gresham's, prepared him solidly in chemistry and physics (to the
exclusion of biology, at least until his last year there when Gresham's

finally put into place a biology curriculum). When it came time for his entrance exams for Cambridge University, he scored firsts in chemistry, physics, and math, but his childhood interests in biology* had prepared him so well for the zoology exam that he scored double firsts in that, outstripping his performance in the other topics. So off to Cambridge he went to pursue an active career in several of the intellectual clubs that are the gravitational attractors of the Cambridge intellectual life. His memberships enabled him to range broadly over the sciences, politics, philosophy, the arts, and economics.

Upon graduating, rather than staying on at Cambridge for his doctoral degree, he decamped for Italy, then to South Africa, where he hoped to study the ecology of the ephemeral lakes and wetlands of that country, the *vleie*, as they are called there in Afrikaans. Clashes over teaching style with his South African professor, H. B. Fantham (who apparently got along with nobody), soon led to Hutchinson being abruptly dismissed from his teaching duties. He was in South Africa on an enforceable contract, though, so after some legal wrangling, it ended happily for him, for he now found himself in the enviable position of being a young man in Africa being paid to do—whatever he wanted. He soon joined his Cambridge colleague and fiancée, Grace Pickford,† who was already in South Africa on her own dime. Together they carried out a collaborative study of the South African wetlands as he had originally wanted to do, but now under the supervision of the more accommodating Lancelot Hogben of the University of Cape Town. There, via a lucky connection through Hogben, Hutchinson learned of a fellowship position at

---

* A famous anecdote has Hutchinson discovering the mummified remains of black rats entombed in the woodwork of the Master's Cottage at Pembroke. Because black rats had long been displaced in the Cambridge area by the Norwegian brown rat, this got Hutchinson thinking about the dynamics of species, which he eventually set down in print. He was eight years old at the time.

† Hutchinson and Pickford married in South Africa, but the marriage lasted only a few years. Pickford went on to a distinguished academic career at the Peabody Museum at Yale.

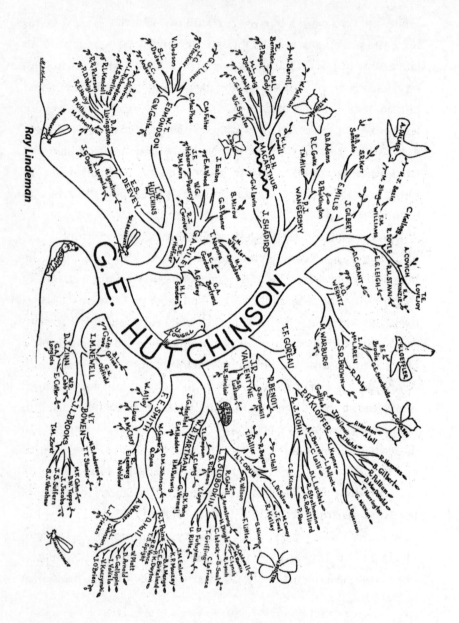

Figure 11.6

Hutchinson's intellectual "family tree." Note Ray Lindeman as a mayfly larva to the left of the base of the tree, and Robert Macarthur (underlined) as the main branch to the left.

Yale, which he failed to get. But his application so impressed the Yale faculty that he was offered another position there, with Alexander Petrunkevitch (August Weismann's last student, as it turns out). This eventually led to Hutchinson's permanent appointment to the Yale faculty, where he remained until his retirement in 1990.

F. Scott Fitzgerald famously described the very rich as just "different from you and me."* On an intellectual level, the observation certainly applies to Hutchinson, who was clearly different from the rest of us. His intellectual career ranged from ecology, to physics, to cybernetics, to anthropology, to behavior, to the classics; and in terms of his intellectual descendants, he left his imprint on virtually the entire scope of twentieth-century biological thought.[3]

Hutchinson's early grounding in chemistry and physics had given him a strong appreciation for the proper place of theory in the practice of science, an appreciation that had not always found a comfortable niche, if I may use that word, in biology. Biologists working in the "real world," that is, working thick into the muck, getting their hands dirty, laboring long hours at the lab bench, inevitably come to a point where they find themselves smacked in the face by life's complexity and diversity, and by what can seem like life's sheer bloodymindedness toward any scientist striving to make sense of it. This experience can easily translate into disdain for the blithe spinner of theory, the same emotion the grizzled military veteran feels for the

---

* The quotation comes from Fitzgerald's 1925 short story "Rich Boy": "Let me tell you about the very rich. They are different from you and me." There is a legendary story that this quotation comes from a conversation between Fitzgerald and Ernest Hemingway, who is said to have responded, "Yes, they have more money." The exchange probably never happened. Hemingway first used the line in a mocking reference to Fitzgerald in the opening of *The Snows of Kilimanjaro,* but Fitzgerald objected to Hemingway's publisher, so the quotation was reattributed to Hemingway's character Julian. The legend of the conversation between Hemingway and Fitzgerald comes from a side note scribbled in one of Fitzgerald's notebooks that referenced "Ernest's wisecrack." In short, the exchange between Fitzgerald and Hemingway never happened. See R. Deis, Quote/Counterquote, "'The rich are different'—a famous quote-counterquote legend," http://www.quotecounterquote.com/2009/11/rich-are-different-famous-quote.html (2009).

rookie just joining the platoon. This was, for many years, an espe-cially difficult problem in ecology, which, of all the biological sci-ences, is probably the one that confronts life in its fullest complexity and diversity. Ecology is also the field where the wreckage of failed theories is piled especially high. Nature in full, it seems, is a capri-cious and brutal muse.

However difficult the relationship between theory and experi-ment might be (or properly between theor*ists*, and experimental-*ists*, for it is very much an issue of personality), when managed well, the tension between them can be extraordinarily fruitful. The physical sciences probably accommodate the tension most effectively, because their subjects are simpler, more cooperative, and tractable. Atoms do not have minds of their own; living things, on the other hand, do.

Nevertheless, Hutchinson sought to bring that same tension to ecol-ogy. This is well illustrated by how he shepherded the pioneering work of one of his early postdoctoral fellows, Raymond Lindeman, through the fraught process of review with the editors of what was then the very practical-minded flagship journal *Ecology*.[4] Lindeman's doctoral work, which he had carried out at the University of Minnesota, was on the flows of energy and matter through a local pond, Cedar Creek Bog. His study wove thermodynamics and food chains into an entirely novel theoretical framework, what he called the *trophic-dynamic* aspect of ecology. As he was writing up his work for publication, Lindeman decamped for Yale to take up a one-year postdoctoral fellowship he had won with Hutchinson. Lindeman's approach to his pond ecosystem resonated with Hutchinson's own conception of ecology, and together they worked up Lindeman's doctoral dissertation for publication, with the first paper slated to go to *Ecology*.

The reviewers picked by *Ecology*'s editor, Thomas Park, emphati-cally rejected Lindeman's paper on the grounds that it would not be sound ecology until Lindeman had done experiments on "fifty or a hundred" more ponds,* and that "desk-produced" papers should not take up "space which should be occupied by research papers."† The sneers of the veteran for the upstart rookie could not have been more clear, and in Lindeman's case, further from the mark. Hutchinson felt compelled to respond even though he was not a coauthor on Lindeman's paper, arguing essentially that biologists needed to adopt a more comfortable relationship with theory, as physicists had done for some time. He argued further that the value of Lindeman's work was not that it established a new "fact" in the "real world," but that it had demonstrated the power of a new way of thinking about the "real world." Thankfully, Hutchinson's brief on Lindeman's behalf carried the day, and after some more editorial wrangling, Lindeman's paper was accepted for publication. Sadly, Lindeman did not live to see his work in print; he died of hepatitis at age twenty-seven, shortly before his paper was published.[5] Today, the trophic-dynamic concept is part of the standard curriculum in ecology.

Lindeman's work is important to our story because it reflected Hutchinson's own conception of the niche as an ongoing transaction of matter and energy between an organism and its environment.[6] In turn, this way of conceiving the niche lent itself to a way of thinking that was similar in its approach to Wright's for the adaptive landscape. Environments could vary along dimensions just as Wright had conceived that traits do. So just as coat color could vary from light to dark, an environmental measure like pond temperature could vary from cool to warm. A dragonfly larva living in that pond could function well or poorly depending on what temperature in that range it inhabited. If good function could be quantified, this could be represented as a curve similar to how Wright plotted fitness against trait.

---

* From the review of Lindeman's paper by Chancey Juday, a prominent limnologist from the University of Wisconsin.

† From the review of Lindeman's paper by Paul Welch of the University of Michigan, another prominent limnologist.

That range of temperatures corresponding to good function could represent a "thermal niche" for the dragonfly larva, that is, that part of the habitat which dragonflies could inhabit and in which they could function well.

Just as Wright could add more and more traits to a fitness landscape, Hutchinson could add other environmental properties to delimit the niche more precisely. One could, for example, plot water salinity as a second niche dimension to pond temperature. There would be some combinations of temperature and salinity where the dragonfly larva functioned well, and others where it would not. Just as Wright had done for his adaptive landscapes, there would be locations on the two-dimensional landscape where the dragonfly larva's "thermohaline"* niche would emerge as high points on that landscape. Again, just as Wright did, more and more dimensions could be added—altitude of the pond, flow-through of pond water, the water's acidity, etc.—as many as we like, and from those, we could construct a niche with as many dimensions as are relevant, all the way up to an n-dimensional hyperniche. Thus was born the Hutchinsonian concept of the niche—one of the most influential concepts in the history of ecology.

Because Hutchinson's n-dimensional niche conception was cast in the same language as Wright's adaptive hyperspace, it was a short leap to merging n-dimensional Darwinian adaptive landscapes with n-dimensional ecological niches, winding all together into an inextricably bound "triple helix" of gene, organism, and environment.[7] This hopeful, beautiful idea offered an escape from the tautology of the apt function gene. Natural selection would drive lineages up to the adaptive peaks of fitness landscape. Those adaptive peaks, in turn, would be tugged into being by the apt functions sitting as the gravitational attractors of the n-dimensional Hutchinsonian niche.

---

* A ten-dollar word that melds temperature (Greek *therme*, heat) and salinity (Greek *hals*, salt).

Closing the circle were the genes that specified the physiological interactions with particular types of environments. Evolution, adaptation, and natural selection would now be driven forward by the relentless turning of the triple-helical screw.

The triple helix metaphor draws additional strength from being testable experimentally. I could draw on many examples to illustrate this, but I will use one that is already familiar: the thermoregulating lizards introduced in Chapter 4. I suggested there that trying to explain behavioral thermoregulation among lizards as the operation of a thermal homeostasis machine—as a clockwork homeostasis— did not adequately account for the essentially cognitive and intentional dimension that homeostasis, properly defined, implies. As it turns out, these very lizards offer a showcase for how Hutchinson's conception of the niche lends itself to an experimental approach. Of course, lizards are far more down-to-earth than an n-dimensional hyperspace, but in this case, that's a feature, not a bug.

Our example will define the thermal niche for a lizard. The definition starts with a lizard's function that is both measurable and ecologically and evolutionarily meaningful for the lizard.[8] Running speed fits that bill well. It's easily measurable: you put a lizard on a runway, goose it in the rear, and measure how fast it sprints away. Running speed is also evolutionarily meaningful. How fast a lizard can run correlates with the likelihood that it can catch prey and avoid being prey itself, both of which feed directly into fecundity and fitness. Finally, running speed is the integrated product of a complex network of presumably gene-specified physiological functions. How fast a muscle contracts, how effectively the heart and lungs can deliver oxygen and fuels to the muscles, how attentive the lizard is to opportunities and dangers—all feed into the ultimate measure of running speed. In turn, these all can be expected to be the product of the natural selection of the genes specifying them.

A lizard's running speed is affected by variation in its temperature

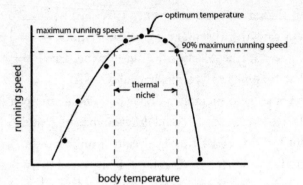

Figure 11.7

Hypothetical running speed performance curve for a lizard. The thermal niche is defined by the temperature range that allows a lizard to run at least 90 percent of its maximum speed.

(Figure 11.7).[9] As every lizard-hunting child knows, cold lizards are sluggish and easy to catch, while warm lizards are canny and elude capture. Graphing out a lizard's running speed versus its temperature gives us a running speed *performance curve:* running speed is slow to nonexistent when the lizard is cold; it rises with temperature up to some maximum speed, and then declines as the lizard gets too hot. The performance curve lets us define a measurable "thermal niche" for the lizard, essentially a Hutchinsonian niche of a single dimension. If we assume that a lizard would do best ecologically if it were able to run faster than, say, 90 percent of its maximum running speed, then we have bracketed the upper and lower bounds of the lizard's thermal niche from its performance curve (Figure 11.7). A lizard would do well to keep its body temperature within that range; that is, if it keeps to its thermal niche.

Adaptation would follow from how well the thermal niche identified by the performance curve matched the availability of those temperatures in the habitat. If suitable habitat temperatures within the thermal niche were rare, the lizard would be poorly adapted to that environment. Selection of apt function genes could presumably shape the performance curve over many generations to bring the

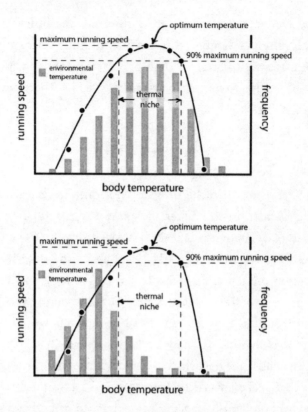

Figure 11.8

Maladaptation (*left*) and adaptation (*right*) in the Hutchinsonian niche. The incidence of environmental temperatures is plotted as a frequency distribution of temperatures (gray bars). When environmental temperatures within the thermal niche are rare (*left*), the lizard is maladapted. When environmental temperatures within the thermal niche are common (*right*), the lizard is well-adapted.

performance curve into line with the habitat's prevailing temperatures (Figure 11.8). Thus, a lizard living, say, in a forested habitat where sun flecks are rare and attaining a high body temperature is difficult or risky might evolve to have a broad thermal niche, that is, a performance curve with high running speeds over a broad range of body temperatures. Presumably, the performance curve would correspond with the need to be able to function at the habitat's commonly available temperatures. This is a sensible solution to the uncertainty that any particular environmental temperature will be available. Better to "spread the risk" and function well over a broad range of temperatures. A lizard living in a sunny habitat where it is easy to attain a high body temperature with little risk might have a narrower performance curve, and therefore a narrow thermal niche.

<hr />

The Hutchinsonian niche provided not only experimental rigor, it laid the foundation for an incredible flowering of theoretical ecology, which is to say mathematical ecology (dare I say, "real" ecology). This efflorescence began in the early 1950s, about the time I was born. I came in on the tail end of it when I was a biology undergraduate at the University of California at Santa Cruz. Although I was a mere larval biologist then, I vividly remember being swept up in the seductive beauty of Hutchinson's "new ecology." Robert MacArthur, perhaps the most brilliant of the primary branches of Hutchinson's intellectual family tree, had died a tragic, premature death from colon cancer just a few years previously, but his legacy burned brightly still—in "the" theory of island biogeography, in the mechanics of niche partitioning among competitors, the subtle behavioral variations that defined the "realized" niches of "MacArthur's warblers"—and we undergraduates couldn't get enough of it. Instead of spending our days in the midst of stuffy museum collections, or sneezing over dusty old volumes in the library hoping to find some nugget of intellectual gold, or slogging through forests swarming with blood-sucking black flies, we spent our time mastering linear algebra and matrix multiplication, sketching 3-D graphs of hypothetical niche spaces, fitness distributions contending with one another—mathematics red in tooth and claw. It was an exhilarating time to be a student.

And yet . . .

Looking back on some of my books from that time, I found this short passage in the introduction to a memorial volume for Robert MacArthur, which captures the spirit of the time, and the problem:

> . . . many ornithologists had recognized that structurally complex habitats like forests harbor more bird species than structurally simple habitats like grassland. However, only when Robert MacArthur devised a simple index of vegetational complexity . . . did it become possible to translate ornithological lore into an equation by which habitat structures are compared and their bird species diversities predicted.[10]

To translate, MacArthur's groundbreaking work on niche partitioning among warblers was the perfect distillation of the Hutchinsonian approach to ecology: melding mathematics and theory with biology to make ecology a "real" science, with the same predictive rigor as other "real" sciences, like physics and chemistry, where a more cordial—or *sensible*—relationship with mathematical abstraction and theory prevailed. What Morgan had done for heredity, what Fisher and Wright had done for population genetics and evolution, now Hutchinson and his many brilliant protégés were doing for ecology: making it a "real" science. All of this is true and admirable, and despite his personal modesty on the question, Hutchinson well deserves the sobriquet assigned to him by his biographer Nancy Slack: the father of *modern* ecology.

Reading that passage nearly forty years later, I am struck by a different perception. Isn't this all a little—I don't know—*Pythagorean*? Look at the passage carefully. We take knowledge—sorry, *lore*—that everyone knew already and turn it into a mathematical model; better yet, a *predictive* mathematical model, that tells us . . . what we already knew. So what, precisely, has actually been gained, besides shrouding knowledge behind the glittering curtains of a mathematical mystery cult?

Which brings us back to Plato Street, for where Pythagoras has raised his head, Plato cannot be far behind. Behind all the intellectual elegance, and all the beautiful mathematics, isn't there a faint odor of the Platonic ideal wafting around the Hutchinsonian niche? The odor has become increasingly strong as the metaphor of the Hutchinsonian niche has been pressed more broadly into service as a heuristic tool, as a metaphor that helps us think about the world. Littered throughout the ecological and evolutionary literature, one can easily find references to such notions as the "predator niche" or the "herbivore niche," or any of the innumerable subdivisions one needs to describe life in the "real" world: the "flying insectivore niche," or the "pelagic filter-feeder niche," or the "grasslands predator niche," or . . . well, you get the idea.[11] If a species of flying insectivore exists, then it exists because a lineage has evolved to "fill" the "flying insectivore niche." If you have a warbler that gleans insects from leaves, it is because a lineage of warblers has evolved to fill the "avian gleaner niche." If one finds two species of flying insectivore, it is because two lineages have divided the "flying insectivore niche" effectively between them: swifts occupy a "diurnal flying insectivore niche," and bats occupy a "nocturnal flying insectivore niche," for example.

The metaphor of the Hutchinsonian niche even provides a handy explanation for the phenomenon of convergence.[12] Success in occupying the "flying insectivore niche" presupposes certain things, such as high maneuverability in flight; ways to range, track, and intercept prey; the means to sweep up the unlucky meal once it's intercepted. And so the lineages of both bats and swifts converge on these necessities, albeit from different starting points. Both have wings and systems of other airfoils that make for high maneuverability while flying. Both have baskets to net their respective prey—fringes of feathers about the mouth for swifts and catcher's mitts of wing membrane for bats. Both use echolocation and sound to track and home in on their prey, and both have brains modified in similar ways to provide the needed computational power for turning sound into landscapes.

This use of the niche metaphor provides satisfying and internally consistent answers, but it ultimately muddies the waters rather than clarifies them. Evolution has now become a striving of organisms toward disembodied and abstract ideals (niches) that draw lineages toward them as strongly as any Platonic demiurge. We may think Darwin purged Plato from evolution, but it seems the old coot keeps sneaking in through the back door.

———◆———

This tendency to crypto-Platonism arises from a presumption about the role the environment plays in the transaction between organism and environment that is the Hutchinsonian niche. The presumption is that the environment is just there, like gravity or sunlight, and that organisms either adapt to it or they don't. From there, it's a short hop to the proposition that *lineages* either adapt to the environment or not. In the "real" world, however, organisms not only draw energy and materials from the environment, they actively remodel the environment to channel matter and energy toward them.[13] As we saw in Chapter 9, earthworms modify soils to help with their water balance, but the tendency extends broadly. Termites build massive lungs out of soil, beavers dam streams to promote the growth of trees they favor as food. These are examples of organisms not just "fitting into" niches, but actively building them. Meld the notion of a "constructed" niche to the Wright adaptive landscape, and you now have an entirely different way to think about adaptation and evolution. Now, the adaptive landscape is under the direct control of the organisms supposedly adapting to it. Now, it becomes an open question: what is adapting to what? Are organisms adapting to the environment, or are environments being made to adapt to organisms?

This way of thinking about adaptation and evolution is called *niche* *construction theory,* and it has been the subject of rather hot debate since its three principal proponents* claimed that it challenges the dominance of what they call "standard evolutionary theory": Neo-Darwinian gene-selectionism, essentially.[14] Niche construction theory rests on some of the themes developed in earlier chapters of this book. How, for example, should hereditary memory be defined? A gene is a memory token that can shape the function of future generations, so it definitely qualifies as hereditary memory. What if the niche constructed by one generation outlasts its lifetime? A beaver pond not only lasts beyond the lifetime of the beaver that built it, but the environmental modifications ramifying from it can last beyond the lifetime of the dam. The work done by a beaver in one generation can therefore be passed on as a hereditary legacy to its descendants—an ecological inheritance that can serve as hereditary memory, just as a gene can.

This broader notion of hereditary memory can extend to a breathtaking array of "constructed niches," ranging from human artifacts and architecture to environments created in the marrow of bones.[15] The challenge of niche construction theory to modern Darwinism is serious, and in my view, compelling. Where the challenge is most strenuous, however, is how niche construction theory admits the phenomenon of agency back into evolutionary thought. And with agency comes homeostasis, intentionality, and purposefulness.[16]

———— • ————

Niche construction theory melds the triple helix of gene, organism, and environment into a coherent theory of adaptation and evolution.

---

* John Odling-Smee of Oxford University, Kevin Laland of St. Andrew's University, and Marcus Feldman of Stanford University. See their *Niche Construction: The Neglected Process in Evolution* (Princeton University Press, 2003).

There remains an unanswered question, though. If evolution proceeds from the relentless turning of the triple helix, what turns the screw?

Here is where agency becomes an unavoidable issue. As mentioned in the Preface, agency can be thought of in various ways. Formally, agency is simply the capacity to act, but within that broad conception, there are different ways to think about it. Consider the question: what is the agent that moves an airplane through the air? In one sense, the engine could be considered the agent, in which case, the agent is simply a machine that does work toward a particular end—in this instance, a flying airplane. On the other hand, the agent moving the airplane could be the pilot, which is a different kind of agency altogether (unless we are prepared to buy into the dubious proposition that the pilot is a mere machine, that is, "meatware" that interfaces with the "hardware" of engine, airfoils, and control systems). The pilot has a desire to maintain the airplane's trim, airspeed, direction, and altitude for the intentional purpose of navigating to a destination. The pilot wants these things in a way the aircraft cannot, even when most of the operations of airplanes are under increasingly sophisticated computer control.[17]

To refer back to the Preface, the first type of agency in the airplane example is the "cumulus cloud" type of agency. In the airplane, the agent is the machinery that channels energy to do the physical work of moving the aircraft forward through the air. The agency of the living pilot can be construed as the alternate "cauliflower" type of agency. Standard evolutionary theory recognizes agency, but it is entirely of the machine type. Genes float about in gene pools, are drawn forward in time by a sort of gravitational force, are sieved through filters as they are drawn upward to the peaks of the adaptive landscapes. What is missing from all of this is *life*—the intentional agency that turns the screw of the triple helix. It is missing not because it is irrelevant, but because it is actively excluded from our thinking. It is epistemic closure, which has tied modern evolution-ism into knots of denial.

To illustrate this problem, let's ask a basic question: why are there flying animals? The trivial answer is because animals have evolved to fly. Beginning about 350 million years ago, at least four lineages of animals have independently evolved the ability to fly under their own power. Insects were the first to launch, sometime during the Carboniferous Period, and were the only animals to fly until the late Triassic Period, roughly 220 million years ago, when the pterosaurs took to the sky. Ten million years after that (or maybe 75 million years after—opinion differs),[*18] another rep-tilian lineage, the birds, took flight. Then the mammals took to the sky, in the form of bats, arising during the Eocene Epoch, about 50 million years ago.

To fly, animals must be equipped to fly. To resurrect Cuvier's pre-Darwinian way of putting it, flying animals must meet the conditions for existence for flight. Today, we would say those ani-mals are adapted to fly. Certainly, all the flying lineages are sim-ilarly equipped to fly: all have airfoils of some sort, and engines to power the animal aloft. Each lineage has gotten there through different paths. The wings of insects, for example, are derived from primitive gills, while the airfoils of bats and pterosaurs are built from sheets of skin that stretch between appendages: between the limbs and body in the case of pterosaurs, or between grotesquely elongated fingers, as in bats. The airfoils of birds are built from many sheetlike feathers, elegantly arranged together into beautiful wings, flaps, and ailerons.

I wish to concentrate on birds, for flight in birds is tied to the birds' distinctive feature, their feathers. The evolution of flight among the

---

* The exact date of the origin of birds is confused and fragmentary. The earliest putative bird fossil is *Proavis*, which seems to have been extant in the late Triassic, 210 million years ago. The famous *Archaeopteryx*, the iconic bird fossil, lived in the late Jurassic, some 75 million years after *Proavis*.

birds seems very closely tied to the evolution of feathers.[19] In the terms of the Hutchinsonian niche, feathers are what enabled birds to occupy the "flying animals niche."

Feathers are one of many forms of growth from the vertebrate skin, which include scales of reptiles, claws, nails, and hair. All are marked by different patterns of folding of the epidermis, which produces outgrowths stiffened from the structural protein keratin. Scales, for example, are sheetlike outgrowths of the epidermis, while hair is a keratin shaft produced by a puckering of the epidermis into the hair follicle. For many years, feathers were thought to be derived from scales, which makes logical sense when one accounts for the similarity of shapes (both are sheetlike in form) and ancestry (feathered birds descended from scaled reptiles). But that turns out to be completely wrong. The contortions in epidermal growth required to produce a feather are entirely different from those that produce a scale. Feathers do not develop from folded sheets, but from a special kind of epidermal puckering that produces a hollow shaft, not a sheet. Feathers are distinctive, even as they are related to the other keratin-based structures that are produced by vertebrate skin.

The conventional story for the evolution of bird flight couples the evolution of the feather with the ability to fly. Thus, a primitive feather could act as a primitive airfoil, which would give a feathered reptile an ever-so-slight selective advantage over animals that were not so blessed. This would be the stepping stone for the next generation, which might have among their number animals with ever-so-slightly better airfoils, conferring upon them a slightly larger selective advantage. And so on and so on. Thus, flight and feathers could have evolved together through a sort of ratcheting selection, culminating in the fully formed airfoil that would allow fully powered flight.

At least three (maybe four, depending on how you count them) scenarios have been proposed for how the parallel and reciprocal evolution of flight and feathers came about.[20] In the "pouncing" scenario, flight evolved from among reptilian sit-and-pounce predators. Here, an animal sits on a prominent outcropping, say, a pile of rocks, and waits for an unsuspecting prey to wander along, at which point, the predator pounces from its promontory and captures its meal. According to the pouncing scenario, reptilian sit-and-pounce predators with proto-feathers could gain slightly more reach and maneuverability as they leapt, and hence more range and accuracy in capturing prey. A related scenario we might call "gliding." Here, flight evolved in animals living in forest canopies, where patrolling, escaping from predators, and defending territories all required the ability to rapidly move between trees without having to travel down to the ground and then back up. The "flying" (or more properly "gliding") squirrels, exemplify this for mammals, and could very well have described tree-dwelling reptilian proto-birds.[21] Proto-birds whose primitive feathers could provide slightly more lift and maneuverability could navigate about the trees slightly more effectively than proto-birds whose feathers could not. Natural selection would then drive the ratchet that would culminate in flight. Then, there is the "running" scenario, where bipedal ground-dwelling dinosaurs (looking something like toothed road runners) were supposedly equipped with fringes of proto-feathers along their forelimbs. These animals would use their feather fringes as nets to capture flying insects as they ran. The sweeping forelimb motions associated with this would produce slight amounts of lift, and with insect-netting leading to ever-more-effective powered hops to catch those insects, powered flight would eventually emerge, as it did on the windy dunes of Kitty Hawk. A variation on this scenario looks to proto-wings to assist in powered leaps up the trunks of trees, reminiscent of Donald O'Connor's incredible acrobatics in his dance number "Make 'Em

Laugh" in the 1952 movie *Singin' in the Rain.** Once off the ground, those animals with proto-feathers could make more of their leap than unfeathered dinosaurs might have been able to.

None of these flight evolution scenarios has ever really gotten very far off the ground. The wings-as-insect-nets idea was particularly problematic, as any feather net that could catch enough air to generate lift would also tend to whoosh insects out of the supposed net, with the end result being a hungry, as opposed to a flying, dinosaur. Similarly, the image of dinosaurs raining from the trees until some learned to fly always had a certain Wile E. Coyote quality about it. What all these scenarios share, though, is the supposition that the evolution of feathers and the evolution of flight somehow went hand-in-hand, as a relentless turning of the triple helix of gene, organism, and environment. We know now that could not have happened. Feathers evolved long before dinosaurs began taking to the sky to occupy the "flying niche."[22] It is no longer possible to entertain any notion of feathers and flight evolving together in a mutual interplay between small adaptations and small increments of function.

So where does that leave adaptation? *Exaptation* is the solution that has been evoked. Exaptations were once called "preadaptations," which are adaptations for one thing that become co-opted to power adaptation to another thing. The evolution of flight provides a good example of such a scenario. Some reptiles have elongated scales that act as umbrellas to keep body temperatures low in

---

* Interestingly, there has been an evolution of sorts of this kind of momentum-propelled flight in the training regimen of *parkour,* also known as free-running. This combines the body's moves with its momentum to propel the body through the air in a kind of proto-"flight." It is claimed that parkour was invented in 1997, but it traces its inspiration to various types of street-dancing moves, which themselves can trace their inspiration back to O'Connor's athletic dancing in *Singin' in the Rain.* There is an amusing variation on the theme of flight and parkour in Nike's "Angry Chicken" commercial (www.youtube.com/watch?v=47tlFVBA130), in which a *traceur* (a practitioner of parkour, in this instance Sébastien Foucan) fails to evade a less-adept chicken, despite his parkour moves.

hot, sunny environments. Elongated scales, therefore, are adaptations to environments that are hot and sunny. Proto-feathers, when they first arose, could have served the same adaptive end, providing shade, but with an umbrella that was constructed differently. Feathers were thus originally adaptations to keep body temperatures cool in hot, sunny environments. However, the hollow-shaft morphology of the proto-feather offered new possibilities for form, as evidenced by the large range of feather forms we see in modern birds, from downy feathers to the elegant airfoils of the wing. The evolution of flight could therefore be the unfolding of the latent adaptive potential inherent in feather growth: feathers were co-opted from an adaptive role in body heat balance to an entirely new form of adaptation, flight. Thus, feathers were an exaptation for flight. Feathers thereby opened up a pathway for the proto-birds to occupy an entirely new Hutchinsonian niche: the "flying niche."

If you think this all sounds like special pleading and scenario spinning, you would be correct. For one thing, the concept of exaptation suffers from the same tautology that afflicts modern Darwinism. Where evolutionary adaptation is selection of genes that promote adaptation, exaptation is an adaptation that leads to another adaptation—it's the same logical fallacy, multiplied. There's also that nagging crypto-Platonism. An exaptation is a new function that opens a trajectory for a lineage to occupy a new Hutchinsonian niche, there waiting to be filled until the right exaptation comes along—the "right" exaptation being, of course, the adaptation that opens up the new evolutionary pathway for the lucky organism possessing it.

And so we are left again tied in knots, necessitated by the need of modern evolutionism to exclude the one thing that could cut through it all: the "cauliflower" type of agency—that form of agency driven by intentionality, striving, purpose, and desire. Could it be that birds fly, not because they were the beneficiaries of lucky exaptations that enabled them to fly, but rather because, in a deep sense, the ancestors of birds *wanted* to fly? They *wanted* to glide from tree to tree, or chase after a tasty lunch, or launch themselves up trunks of trees to avoid being lunch themselves. And those wants have dragged the genes into the future in their tumultuous intentional wake. And this makes evolution at root a phenomenon of cognition, of intentionality, of purpose, of desire—of *homeostasis*.

At this point Thomas Hunt Morgan's ghost can be counted upon to enter the room and begin rattling his chains, warning us that we cannot speak of such things and hope to remain scientists. To invoke intangibles such as wants, desires, and motivations is to yield to the V-word—vitalism. Let us grant that Morgan's ghost has a point. Yet when he drifts from the room and glides on to torment others thinking their forbidden vitalist thoughts, the dilemma that has animated this book, indeed that haunts all of modern biology, does not go away: what if purpose and desire are *the* fundamental attributes of anything that lives, and everything about them?

As I have tried to show throughout this book, modern biology is burdened heavily by avoidance of that question. The avoidance has not been driven by any of the usual motivators: fear, procrastination, denial, evasion of an unpleasant truth. Rather, the motivation has had at its heart the utmost idealism: to establish a true science of life and a true science of evolution that could refute any temptation to relegate the science of life to the realm of stamp collecting. No matter the motivation, though, biology in general, and evolutionism in particular, has largely been led astray by that quest. Where we have striven to exclude the ghosts from our machines, we have inadvertently constructed back doors, like Hutchinson's niche, that allow the ghosts to creep right back in. Where we have striven to tame life and evolution to mathematical rigor, we have only ended up restating obvious truths, but now hidden from all but a limited set of mathematical devotees. There is a missing ingredient, it seems, and that ingredient is the living organism that is marked by purpose and desire. Without that missing ingredient, modern evolution is just a magnificent contrivance, the wooden bird in the cuckoo clock that is no more alive than the cogs and springs and bellows that move it.

EPILOGUE

Evolution, Purpose, and Desire

Richard Dawkins has famously been quoted as saying that Darwinism made it possible for him to be an intellectually fulfilled atheist. Even if I cannot share the conclusion, I can respect the sentiment. It's impossible for anyone who has a heart capable of wonder not to be struck with awe when contemplating the living world. Living nature is a wonder to behold, and it's a wonder to witness the energy and joy of those who seek to understand it. When you find a convincing explanation for it all, it's only natural to be filled with the peace that comes with understanding. I can grant that sense of fulfillment to Dawkins, even if I think he's wrong-headed about its source.

Here's the problem: the voice that beckons Richard Dawkins to fulfillment is his own—it is not nature's. The edifice of modern Darwinism, as magnificent an edifice as the most beautiful cathedral, an edifice as painstakingly built by the work of generations of devoted and skilled artisans, is hollow at its core. It is an echo chamber. What modern Darwinism is asking us to admire is a husk of something once living, but with its vital core drained away as we have poked and prodded with our naughty thumbs until we are left with nothing but the beautiful shell. In short, the science of life has become disenchanted with life itself. That is the looming crisis I described in the Preface.

The crisis will be averted, I assert, when biology becomes reenchanted. How, or even whether, this happens is anybody's guess, but my candidate for the reenchanting of biology is Claude Bernard's dangerous idea: homeostasis, the relentless striving of living systems for persistence and self-sustenance. Properly understood, homeostasis is life's fundamental property, what distinguishes it from non-life. In short, homeostasis *is* life. It is a first principle that stands on its own and does not derive from any process associated with life, including its evolution.

This may be a hard pill to swallow, because it upends most of the magnificent edifice of modern Darwinism. Now, homeostasis does not derive from natural selection; it is homeostasis that drives selection, and there is nothing natural about it. What drives the course of evolution is not the soulless lottery of the gene pool, but life's striving for persistence. The striving is driven not by the luck of the lottery, but by a cognitive sense of self, even down to the smallest bacterium, even preceding, as I have argued, the emergence of life itself. A deep intelligence is at work in life, its operations, and its history, and it cannot be denied. Yet that is precisely what modern Darwinism asks us to do. So, I shall make a counterclaim to Dawkins's: homeostasis, properly understood as the radical idea it is (Bernardism if you will), makes it possible to be an intellectually fulfilled vitalist.

———•———

But, so what if it can do this? Science, including the science of life, has been successfully ticking along for so long in the world of mechanism and materialism that it seems to be the new normal: there seems to be no other imaginable way to think about life. True enough, the enterprise of science continues to enjoy generous funding, support, and prestige. Biology makes dazzling advances almost daily in new products and therapies. It continues to challenge our views of our world, ourselves, and our society. So where's the crisis? Why change?

Here is my claim for why. Science is first and foremost a cultural phenomenon. Its objective value comes from its practice of querying nature itself for answers to what nature is. This is what makes science distinctive as a philosophy of nature. I don't think I would find much disagreement on this point. Where things begin to get dicey is the relationship of science with the broader culture: no matter how ardently it is desired, science cannot really hold itself apart from the culture in which it is embedded. This is not a claim for dominance of culture over science. In the best liberal tradition, science can be a powerful voice to shape culture, but there is no escaping that science is also shaped by culture. The crisis for biology right now is one of alienation: of the alienation of the science of life from life itself, but more alarmingly, of the alienation of science from the broader culture. Evolution has been the touchpoint for this alienation for a very long time. It would be nice if it could be resolved.

A few years ago, while driving from Austin, Texas, to our home near Syracuse, New York, our route took us within a few miles of Dayton, Tennessee, site of the 1925 Scopes Trial, where John T. Scopes was prosecuted for teaching evolution in his classroom. We were in a little bit of a rush—we were trying to outrace a snowstorm barreling across the Midwest that would have made our travel home difficult—but it was a short detour, and it was a beautiful crisp winter day, so we took a diversion to visit the town.

Dayton is not quite like it was in the 1920s—you approach it through the usual gauntlet of ugly strip malls, dollar stores, and car dealerships. Yet the downtown itself, where the Rhea County

The Rhea County Courthouse in Dayton, Tennessee, and statue of William Jennings Bryan. The statue of the "Great Commoner" can be seen to the left of the courthouse.

Courthouse still stands, retains much of its southern charm. The courthouse is still a working courthouse. We were not allowed to look in on the Scopes courtroom during our short visit because it was occupied with a trial, and we watched from the spacious grounds as prisoners were being offloaded for their own appointments with judicial destiny. The site still spoke volumes, though. On the grounds, for example, is an imposing statue of William Jennings Bryan, who prosecuted the Scopes case, and who lends his name to Bryan College, a local Christian college (see figure). Clarence Darrow, who defended Scopes, does not have a statue. Dayton's culture has made its choice clear for which of the two, Bryan or Darrow, was the hero of that trial.

Are the good citizens of Dayton wrong in their choice? Are they the uneducated rubes that H. L. Mencken, caustic chronicler of the Scopes Trial, painted them to be? Was John Scopes our American Socrates, the victim of a conflict between intolerant religious faith and the brave spirit of rational inquiry, as we might believe if *Inherit the Wind* were our only guide to the case?* Was it a circus of civic boosterism blown out of all proportion?[1] Looked at objectively, the Scopes case was a little bit of all these things, but mostly it was a proxy for a larger fight over who controls the culture. Can "science" claim supremacy just by virtue of it being "science"? Or does "science" have to take its place on an equal footing with other claimants, including those who might rank science rather low on their list of cultural priorities? The question is a big one that goes right to the heart of freedom of thought and freedom of expression. In terms of defending freedom of thought, it was actually Bryan, and not Darrow, who probably had the stronger argument.[2] It seems we still have something to learn from the Scopes case—and of the complex interaction between science and culture. Can we really hold them apart?

Recent history suggests we have learned little from these lessons, if for no other reason than we keep litigating the same issue over and over, from Tennessee to Arkansas to California to Louisiana to Texas to Dover, Pennsylvania.[3] The lawyers can pick through the legal details of these cases all they want, but they all revolve around the same tedious bone of contention that animated the Scopes trial: who shall determine the terms of the culture?—"science" and "enlightenment," or the ignorant hoi polloi? Have we really come no further?

There are no pat answers to these questions, of course. On the one hand, one looks at the ongoing "scientific" questions that have roiled our culture, from the creation wars through eugenics through genetic engineering through climate change to intelligent design theory, and one despairs at the polarization, politicization, and demonization that accompanies these controversies. Is that the best we can expect from what is ostensibly an attempt to understand

---

* The 1955 play by Jerome Lawrence and Robert Edwin Lee (Ballantine, 2003), which has also seen four film adaptations, starring the likes of Spencer Tracy, Gene Kelly, Jason Robards, Kirk Douglas, Jack Lemmon, and George C. Scott.

nature in full? Fortunately, hope still floats because one can see difficult controversies negotiated in a climate of the utmost respect and
mutual goodwill, as in sorting through the difficult ethical issues
surrounding research with fetal stem cells.[4] Sometimes, hope floats
along on hidden reservoirs of the best tradition of classical liberal-
ism, as in Stephen Jay Gould's quiet supervision of a young Earth
creationist, Kurt Wise, for his doctoral degree in paleontology.[5] And
on the *other* hand, as Tevye might say,* despair rises again when one
considers the shabby treatment that has been meted out to various
advocates of and sympathizers to intelligent design theory, even
to academicians with long-standing solid reputations who suggest
there is a legitimate critique to be made of Darwinism, at least in
its modern form.[6] Science thus seems rather delicately poised on the
cusp of its relationship to culture. Which way will we go? Alienation
or accommodation?

The dilemma was illustrated for me in a memorable scene in the
film adaptation (by Josh Boone) of John Green's 2012 novel *The Fault
in Our Stars*. In the novel, a young cancer patient, Hazel Lancaster,
and her boyfriend, Augustus Waters (Gus),† also a cancer patient,
travel to Amsterdam to visit the author of a book that has capti-
vated both of their imaginations. The book, *An Imperial Affliction*,
is about childhood cancer, and they opened a correspondence with
the book's reclusive author, Peter van Houten, seeking resolution
to the novel's ambiguous ending (it ends in mid-sentence, presumably with the death of the book's protagonist). To their surprise, van
Houten responds with an invitation to come to Amsterdam if they
want answers to their questions. That they do, but their meeting

---

* Tevye was the central character in Norman Jewison's 1971 film, *Fiddler on the Roof*, based upon
Sholem Aleichem 1905 novel *Tevye and His Daughters*. Tevye, played by Chaim Topol in the film,
was torn between the demands of tradition and new mores of the future. He struggled with these
through soliloquies marked by "On the one hand . . . , but on the other hand . . ."

† Hazel is played by Shailene Woodley and Gus by Ansel Elgort.

with van Houten (played to craggy perfection by Willem Dafoe) is a disaster. The encouraging correspondence turns out to have been not with van Houten, but with van Houten's good-natured assistant, Lidewij. Van Houten himself turns out to be an abusive, rage-filled alcoholic who ends up berating his visitors with a rant—that the two seekers in front of him are nothing but bags of atoms and that their afflictions are nothing but an accident of natural selection, "a failed experiment in mutation," with no more meaning to anyone than a cloud passing in the sky.

Much of the movie hovers on the edge of the maudlin, but that scene is powerful with brutal honesty, driving home a troubling point about the place that science, and the science of evolution in particular, has come to assume in the cultural ethos of our day. We scientists presume to be the custodians of a superior way of thinking about the universe and everything in it. Yet if that scene is any indication of the broader culture's opinion of our presumptions, we have lost the argument. We may retain standing in the public square, but it is not based upon our appealing philosophy. Rather, it is based upon our ability to deliver goods: better pharmaceuticals, better treatments for dread diseases, better ways to manage our environment, better ways to get us around the planet. As a philosophy of nature, though, as a shaper of our culture as we strive for it to be, science has come to be widely looked upon with suspicious dread. That van Houten couched his rant in the nihilistic language of modern Darwinism underscores the point. That Green put those words in the mouth of a shriveled, irrelevant, isolated, drunken has-been underscores the point painfully.

Peter van Houten is a caricature, of course, but Green's novel poses, in an interesting way, the problem that biology, and evolutionism, now faces. The story arc of *The Fault in Our Stars* follows Hazel and Gus as they chart their difficult path between two radically opposing poles: the mindless religiosity of the church support group where they met, and Peter van Houten's desiccated nihilism. The dilemma is similar in many ways to the dilemma biology faces today. For nearly a century, our choice has been stark: the purposeless world of the materialist, or the demon-haunted world of the vitalist. For nearly a century, we have been forced to choose, and casting your lot with one has meant being cast out from the other. But there is a middle path to follow, which I have argued in this book means coming to grips with life's truly distinct nature—its purposefulness, its intentionality, and its distinctive intelligence. Failing to do this will only cast us deeper into the shadows of irrelevance.

ACKNOWLEDGMENTS

I owe many tremendous debts of thanks to many people. The initial writing and research for the book was undertaken during a sabbatical leave in 2010 at St. John's College, Cambridge University. I was hosted there by the inimitable Professor Simon Conway Morris, who introduced me to the Fellow's Table and other pleasures of the Cambridge academic life. My time at Cambridge was made possible through the generous support of a fellowship from the John Templeton Foundation, which has ever since patiently awaited the book that I promised to deliver seven years ago. I'm not sure which I am more grateful for: the support itself, or the leeway the foundation has given me to take the time to think the book through thoroughly. The Stellenbosch Institute for Advanced Study in South Africa generously provided me an academic home during the final stages of shepherding this manuscript to publication, which included the opportunity for extended and illuminating conversations with Addy Pross, whom I introduced in Chapters 2 and 10. I am also grateful to the friends and colleagues upon whom I have imposed to read the book and give me the feedback I need, including Berry Pinshow of Ben-Gurion University of the Negev; Zann Gill of the Microbes and Mind Forum; Paul Bardunias, my postdoc, who gave me useful insights into the history of warfare; David Morris of the McGregor Museum in Kimberley, South Africa, who offered his thoughts at some crucial points in the manuscript; my friend Eugene Marais of the National Museum of Namibia, whom you met in Chapter 1; and Lisa Margonelli, the author of the engaging book *Oil on the Brain,* and who, against all advice that is wise, is now writing a popular book about termites.

I owe special thanks to my long-suffering family: my wife, Debbie, who offered wise advice on how to grab readers and hold them and who stands by me even if I am often off in the ether; and my daughters, Jackie and Emma, both grown now, for the simple pleasure of their existence.

Finally, I thank several individuals who have helped bring this book to fruition. My agent, Roger Freet of Foundry Literary and Media, helped negotiate the publication. Thank you, Roger. Stephen Meyer of the Discovery Institute helped guide me and my book to Roger Freet, an act of generosity for which I am very grateful. My editor at HarperOne, Mickey Maudlin, has improved the manuscript in innumerable and superlative ways, mostly by encouraging me to stop and take a breath every so often. I am also grateful to Anna Paustenbach, editorial assistant to Mickey Maudlin at HarperOne, who along with Mickey helped shape the title.

Apropos of the title, I wish to offer a short comment on that. *Purpose and Desire* is intended as a counterfoil to Jacques Monod's luminous and influential 1970 book *Chance and Necessity.* That book has cemented into place, perhaps more than any other of the modern era, the notion that we biologists must make the Hobson's choice if we are to be scientists—for that matter, as must anyone who claims a scientific or even an avocational interest in the phenomenon of life. It's a beautiful and deep book, as would be expected from such an author as Monod, but its thesis rests upon what I have come to believe is a distorted picture of the history and philosophy of the science of life. In Monod's case, the picture he painted was shaped by the author's having already made the Hobson's choice himself and being happy with his decision. I, too, made the Hobson's choice early in my scientific career, and it was the conventional choice you might imagine: deeply reductionist, materialist, and mechanist. Over the past few years, however, I have been rethinking that choice, and I have come to the conclusion that I didn't choose entirely correctly. It is for that reason that I sometimes describe myself as a recovering reductionist. For what it's worth, I offer this book as an outline of how I came to change my mind. I hope that spirit came through as you read it.

Scott Turner
Tully, New York

NOTES

Preface

1 F. Crick, *Astonishing Hypothesis: The Scientific Search for the Soul* (Scribner, 1995), 3.

2 R. Dawkins, *River Out of Eden: A Darwinian View* (Perseus, 2008), 133.

3 D. C. Dennett, *Darwin's Dangerous Idea: Evolution and the Meanings of Life* (Simon & Schuster, 1995), 59.

4 C. Sagan, *The Demon-Haunted World: Science as a Candle in the Dark* (Ballantine, 1997).

Chapter 1: The Pony Under the Tree

1 M. Rose and L. Mueller, *Evolution and Ecology of the Organism* (Benjamin Cummings, 2006), 66–67.

Chapter 2: Biology's Second Law

1 C. D. Moyes and P. M. Schulte, *Principles of Animal Physiology,* 2nd ed. (Benjamin Cummings, 2008).

2 K. Schmidt-Nielsen, *How Animals Work* (Cambridge Univ. Press, 1972).

3 D. C. Dennett, *Darwin's Dangerous Idea: Evolution and the Meanings of Life* (Simon & Schuster, 1995).

Chapter 3: Many Little Lives

1 H. Bloch, "François Magendie, Claude Bernard, and the Inter-Relationship of Science, History and Philosophy," *Southern Medical Journal* 82 (1989): 1259–1261.

2 M. Foster, *Claude Bernard* (Longman's Green, 1899).

3 C. G. Gross, "Claude Bernard and the Constancy of the Internal Environment," *Neuroscientist* 4 (2008): 380–385.

4 Bloch, "François Magendie."

5 S. Normandin, "Claude Bernard and an Introduction to the Study of Experimental Medicine: 'Physical Vitalism,' Dialectic, and Epistemology, *Journal of the History of Medicine and Allied Sciences* 62 (2007): 495–528, doi:10.1093/jhmas /jrm015; C. T. Wolfe, "Introduction: Vitalism without Metaphysics? Medical Vitalism in the Enlightenment," *Science in Context* 21 (2008): 461–463; T. Cheung, "Regulatory Agents, Functional Interactions, and Stimulus-Reaction Schemes: The Concept of 'Organism' in the Organic System Theories of Stahl, Bordeu and Berthoz," *Science in Context* 21 (2008): 495–519.

6 A. Petit, "Claude Bernard and the History of Science," *Isis* 78 (1987): 201–219, doi:10.1086/354390.

7 J. Longrigg, "Presocratic Philosophy and Hippocratic Medicine," *History of Science* 27 (1989): 1–39, doi:10.1177/007327538902700101.

8 R. G. DePalma, V. W. Hayes, and L. R. Zacharski, "Bloodletting: Past and Present," *Journal of the American College of Surgeons* 205 (2007): 132–144, doi:http: //dx.doi.org/10.1016/j.jamcollsurg.2007.01.071; N. W. Kasting, "A Rationale for Centuries of Therapeutic Bloodletting: Antipyretic Therapy for Febrile Diseases," *Perspectives in Biology and Medicine* 33 (1990): 509–516; I. H. Kerridge and M. Lowe, "Bloodletting: The Story of a Therapeutic Technique," *Medical Journal of Australia* 163 (1995): 631; J. H. Warner, "Therapeutic Explanation and the Edinburgh Bloodletting Controversy: Two Perspectives on the Medical Meaning of Science in the Mid-Nineteenth Century," *Medical History* 24 (1980): 241–258.

9 Normandin, "Physical Vitalism," and Cheung, "Regulatory Agents."

10 Wolfe, "Vitalism without Metaphysics?"

11 Cheung, "Regulatory Agents."

12 P. M. Dawson, *A Biography of François Magendie* (Albert T. Huntington, 1908).

13 Magendie (1819) quoted in Dawson, *Biography of François Magendie*, 6–7.

14 C. Bernard, *Experimental Medicine*, trans. Henry Copley Greene (Transaction, 1999), 88–89.

15 Normandin, "Physical Vitalism."

Chapter 4: A Clockwork Homeostasis

1 A. Gottlieb, *The Dream of Reason: A History of Western Philosophy from the Greeks to the Renaissance* (Norton, 2000).

2 L. Menand, *The Metaphysical Club: A Story of Ideas in America* (Farrar, Straus and Giroux, 2002); H. Hein, "The Endurance of the Mechanism: Vitalism Controversy," *Journal of the History of Biology* 5 (1972): 159–188, doi:10.1007/bf 02113490.

3 See, for example, P. Cohen in "'Epistemic Closure'? Those Are Fighting Words,

Friend," New York Times (April 27, 2010): C1; www.nytimes.com/2010/04/28/books/28conserv.html.

4 M. Ware and M. Mabe, *The STM Report: An Overview of Scientific and Scholarly Journal Publishing*, 3rd ed. (International Association of Scientific, Technical and Medical Publishers, 2012).

5 E. Humes, *Monkey Girl: Evolution, Education, Religion, and the Battle for America's Soul* (HarperCollins, 2009); R. L. Numbers, ed., *Galileo Goes to Jail, and Other Myths about Science and Religion* (Harvard Univ. Press, 2009); R. Lewin, "Judge's Ruling Hits Hard at Creationism," *Science* 215 (1982): 381–384; J. S. Turner, "Signs of Design," *Christian Century* 124 (2007): 18–22; T. Nagel, *Mind and Cosmos: Why the Materialist Neo-Darwinian Conception of Nature Is Almost Certainly False* (Oxford Univ. Press, 2012).

6 A. L. Hughes, "The Folly of Scientism," *New Atlantis* 37 (2012): 32–50.

7 F. A. Hayek, "The Intellectuals and Socialism," *University of Chicago Law Review*, Spring 1949, 417–423, 425–433.

8 E. Mayr, *The Growth of Biological Thought: Diversity, Evolution, and Inheritance* (Belknap/Harvard Univ. Press, 1982); E. Mayr, *What Makes Biology Unique?: Considerations on the Autonomy of a Scientific Discipline* (Cambridge Univ. Press, 2007).

9 L. J. Henderson, *The Fitness of the Environment: An Inquiry into the Biological Significance of the Properties of Matter* (MacMillan, 1913).

10 W. B. Cannon, "Physiologic Regulation of Normal States: Some Tentative Postulates concerning Biological Homeostatics," in *Jubilee Volume to Charles Richet* (Éditions Medicales, 1926), 91–93.

11 D. Ullman, "The Philosophical and Historical Roots of the Holistic Approaches to Health: A Review of the First Three Volumes of 'Divided Legacy' by Harris L. Coulter, PhD," (Homeopathic Educational Services, 2007); also available at https://www.homeopathic.com/Articles/Introduction_to_Homeopathy/A_review_of_the_first_three_volumes_of_Divid.html.

12 V. D. Hanson, *The Savior Generals: How Five Great Commanders Saved Wars That Were Lost—From Ancient Greece to Iraq* (Bloomsbury, 2014).

13 W. R. Murray and A. R. Millett, *Military Innovation in the Interwar Period* (Cambridge Univ. Press, 1998); T. Moy, *War Machines: Transforming Technologies in the U.S. Military, 1920–1940* (Texas A&M Univ. Press, 2001).

14 S. J. Heims, *John von Neumann and Norbert Wiener: From Mathematics to the Technologies of Life and Death* (MIT Press, 1982).

15 N. Wiener, "Men, Machines, and the World About," in *Medicine and Science*, ed. I. Gakderston, (International Universities Press, 1954), 13–28.

16 D. A. Mindell, "Anti-Aircraft Fire Control and the Development of Integrated Systems at Sperry, 1925–40," *IEEE Control Systems* 15 (1995): 108–113,

doi:10.1109/37.375318; D. A. Mindell, "Automation's Finest Hour: Bell Labs and Automatic Control in World War II," *IEEE Control Systems* 15 (1995): 72, doi:10.1109/MCS.1995.476388; M. Gladwell, TED, "The Strange Tale of the Norden Bombsight," https://www.ted.com/talks/malcolm_gladwell?language =en (2011); A. L. Pardini, *The Legendary Norden Bombsight* (Schiffer, 1999).

17 A. Rosenblueth, N. Wiener, and J. Bigelow, "Behavior, Purpose and Teleology," *Philosophy of Science* 10 (1943): 18–24.

18 S. A. Umpleby, "A History of the Cybernetics Movement in the United States," *Journal of the Washington Academy of Sciences* 91 (2005): 54–66.

19 D. Stanley-Jones and K. Stanley-Jones, *The Kybernetics of Natural Systems: A Study in Patterns of Control* (Pergamon, 1960).

20 Skinner quoted in "Science and the Citizen: Paradise of 1984," *Scientific American* 196 (1957): 58–60.

21 O. Lippold, "Physiological Tremor," *Scientific American* 224 (1971): 65–73; A. Benabid, P. Pollak, M. Hommel, J. M. Gaio, J. de Rougemont, and J. Perret, ["Treatment of Parkinson Tremor by Chronic Stimulation of the Ventral Intermediate Nucleus of the Thalamus"], *Revue neurologique* 145 (1988): 320–323.

22 R. W. Mann, "Cybernetic Limb Prosthesis: The ALZA Distinguished Lecture," *Annals of Biomedical Engineering* 9 (1981): 1–43; M. C. Carrozza, Paolo Dario, Fabrizio Vecchi, and F. Sebastiani, "The CyberHand: On the Design of a Cybernetic Prosthetic Hand Intended to Be Interfaced to the Peripheral Nervous System," in *Intelligent Robots and Systems, 2003 (IROS 2003). Proceedings* (IEEE, 2003), 2642–2647.

23 E. F. Du Bois, *Lane Medical Lectures: The Mechanism of Heat Loss and Temperature Regulation*, Vol. III (Stanford Univ. Press, 1937); H. Hensel, "Neural Processes in Thermoregulation, *Physiological Reviews* 53 (1973): 948–1017.

24 M. J. Kluger and L. K. Vaughan, "Fever and Survival in Rabbits Infected with *Pasteurella multocida, Journal of Physiology* 282 (1978): 243–251; J. T. Stitt, "Fever vs. Hyperthermia," *Federation Proceedings* 38 (1979): 39–43.

25 K. Nagashima, S. Nakai, M. Tanaka, and K. Kanosue, "Neuronal Circuitries Involved in Thermoregulation," *Autonomic Neuroscience* 85 (2000): 18–25, doi:10.1016/s1566–0702(00)00216–2.

26 A. van der Schoot, "Kepler's Search for Form and Proportion," *Renaissance Studies* 15 (2001): 59–78.

27 J. W. Hudson, "Torpidity in Mammals," in *Comparative Physiology of Thermoregulation*, Vol. III: *Several Aspects of Thermoregulation*, ed. G. C. Whittow (Elsevier, 1973), 98–166; J. R. King and D. S. Farner, "Energy Metabolism, Thermoregulation and Body Temperature," in *Biology and Comparative Physiology of Birds*, Vol. II, ed. A. J. Marshall (Elsevier, 1961), 215–288; S. Kobayashi, "Temperature-Sensitive Neurons in the Hypothalamus: A New Hypothesis That

They Act as Thermostats, Not as Transducers," *Progress in Neurobiology* 32 (1989): 103–135; F. Kronenberg and H. Heller, "Colonial Thermoregulation in Honeybees (Apis mellifera)," *Journal of Comparative Physiology* 148 (1982): 65–76; C. L. Prosser, "Temperature," in *Comparative Animal Physiology,* ed. C. L. Prosser (Thomson Learning, 1973), 362–428; A. J. Romanovsky, "Thermoregulation: Some Concepts Have Changed. Functional Architecture of the Thermoregulatory System," *American Journal of Physiology: Regulatory, Integrative and Comparative Physiology* 292 (2007): R37–R46, doi:10.1152/ajpregu.00668.2006; J. R. Templeton, "Reptiles," *Comparative Physiology of Thermoregulation* (1970): 167–221.

28 M. Kleiber, *The Fire of Life: An Introduction to Animal Energetics* (R. E. Krieger, 1975).

29 M. Cabanac, "Sensory Pleasure," *Quarterly Review of Biology* 54 (1979): 1–29.

30 R. B. Cowles and C. M. Bogert, "A Preliminary Study of the Thermal Requirements of Desert Reptiles," *Bulletin of the American Museum of Natural History* 83 (1944): 261–296; M. L. Berk and J. E. Heath, "An Analysis of Behavioral Thermoregulation in the Lizard *Dipsosaurus dorsalis,*" *Journal of Thermal Biology* 1 (1975): 15–22.

31 H. Dreisig, "Control of Body Temperature in Shuttling Ectotherms," *Journal of Thermal Biology* 9 (1984): 229–234; R. F. Hainsworth, "Optimal Body Temperatures with Shuttling: Desert Antelope Ground Squirrels," *Animal Behaviour* 49 (1995): 107–116, doi:http://dx.doi.org/10.1016/0003-3472(95) 80158-8.

32 M. L. Berk and J. E. Heath, "Effects of Pre-optic, Hypothalamic and Telencephalic Lesions on Thermoregulation in the Lizard, *Dipsosaurus dorsalis,*" *Journal of Thermal Biology* 1 (1975): 65–78.

33 H. A. Bernheim and M. J. Kluger, "Fever and Antipyresis in the Lizard *Dipsosaurus dorsalis,*" *American Journal of Physiology* 231 (1976): 198–203.

34 M. J. Kluger, "Phylogeny of Fever, *Federation Proceedings* 38 (1979): 30–34.

35 R. B. Huey and T. P. Weber, "Thermal Biology of *Anolis* Lizards in a Complex Fauna: The *Cristatellus* Group on Puerto Rico," *Ecology* 57 (1976): 985–994.

36 R. B. Huey and M. Slatkin, "Costs and Benefits of Lizard Thermoregulation. *Quarterly Review of Biology* 51 (1976): 363–384.

37 P. E. Hertz, R. B. Huey, and E. Nevo, "Fight vs. Flight: Thermal Dependence of Defensive Behavior. *Animal Behavior* 30 (1982): 676–679; R. B. Huey and T. Preston, "Thermal Biology of a Solitary Lizard: *Anolis marmoratus* of Guadeloupe, Lesser Antilles," *Ecology* 56 (1975): 445–452; P. E. Hertz, R. B. Huey, and E. Nevo, "Homage to Santa Anita: Thermal Sensitivity of Sprint Speed in Agamid Lizards," *Evolution* 37 (1983): 1075–1084.

38 P. E. Hertz and E. Nevo, "Thermal Biology of Four Israeli Agamid Lizards in Early Summer," *Israel Journal of Zoology* 30 (1981): 190–210.

39 W. T. Powers, *Behavior: The Control of Perception* (Benchmark, 2005).

Chapter 5: A Mad Dream

1 M. T. Ghiselin, "The Imaginary Lamarck: A Look at Bogus 'History' in Schoolbooks, *Textbook Letter* (September–October 1994), http://www.textbookleague.org/54marck.htm.

2 C. C. Gillispie, "Lamarck and Darwin in the History of Science," *American Scientist* 46 (1958): 388–409.

3 Gillispie, "Lamarck and Darwin."

4 J. E. Reiss, *Not by Design: Retiring Darwin's Watchmaker* (Univ. of California Press, 2009); R. W. J. Burkhardt, *The Spirit of System: Lamarck and Evolutionary Biology* (Harvard Univ. Press, 1995).

5 D. W. McShea and R. N. Brandon, *Biology's First Law: The Tendency for Diversity and Complexity to Increase in Evolutionary Systems* (Univ. of Chicago Press, 2010); R. E. Morel and G. Fleck, "A Fourth Law of Thermodynamics," *Chemistry* 15 (2006): 305–310; R. Swenson, "The Fourth Law of Thermodynamics or the Law of Maximum Entropy Production," *Chemistry* 18 (2009): 333–339.

6 S. J. Gould, "Is a New and General Theory of Evolution Emerging?" *Paleobiology* 6 (1980): 119–130; S. J. Gould and R. C. Lewontin, "The Spandrels of San Marco and the Panglossian Paradigm: A Critique of the Adaptationist Programme," *Proceedings of the Royal Society of London* 205B (1979): 581–598.

7 L. W. Alvarez, W. Alvarez, F. Asaro, and H. V. Michel, "Extraterrestrial Cause for the Cretaceous-Tertiary Extinction," *Science* 208 (1980): 1095–1108; W. Alvarez, E. G. Kauffman, F. Surlyk, L. W. Alvarez, F. Asaro, and H. V. Michel, "Impact Theory of Mass Extinctions and the Invertebrate Fossil Record," *Science* 223 (1984): 1135–1141.

8 D. W. Goldsmith, *Nemesis: The Death-Star and Other Theories of Mass Extinction* (New York: Walker, 1985); C. W. Harper Jr., "Might Occam's Canon Explode the Death Star?: A Moving-Average Model of Biotic Extinctions," *Palaios* 2 (1987): 600–604; D. M. Raup and J. J. Sepkoski, "Periodicity of Extinctions in the Geologic Past," *Proceedings of the National Academy of Sciences of the United States of America* 81 (1984): 801–805.

9 Reiss, *Not by Design.*

10 K. S. Thomson, "Natural Theology," *American Scientist* 85 (1997): 219–221; W. Paley, *Natural Theology* (Bobbs-Merrill, 1802).

11 A. Pattison, *The Darwins of Shrewsbury* (History Press, 2009).

12 A. Guerrini, "Archibald Pitcairne and Newtonian Medicine," *Medical History* 31 (1987): 70–83.

13 E. Darwin, *Zoonomia; or The Laws of Organic Life,* Vol. 1 (1794), 1.

14 A. R. Wallace, "On the Tendency of Varieties to Depart Indefinitely from the Original Type (S43: 1858). *Journal of the Proceedings of the Linnean Society: Zoology* 3 (1858): 45–62.

15 Gillispie, "Lamarck and Darwin."

16 C. Darwin, *The Variation of Animals and Plants under Domestication* (J. Murray, 1868).

17 W. F. McComas, "Darwin's Invention: Inheritance and the 'Mad Dream' of Pangenesis," *American Biology Teacher* 74 (2012): 86–91, doi:10.1525/abt.2012.74.2.5.

18 R. W. Burkhardt, "Closing the Door on Lord Morton's Mare: The Rise and Fall of Telegony," *Studies in the History of Biology* 3 (1979): 1–21.

19 W. F. McComas, "Darwin's Invention."

20 G. S. Dawe, X. W. Tan, and Z.-C. Xiao, "Cell Migration from Baby to Mother," *Cell Adhesion & Migration* 1 (2007): 19–27; Y. Liu, "A New Perspective on Darwin's Pangenesis," *Biological Reviews* 83 (2008): 141–149, doi:10.1111/j.1469 –185X.2008.00036.x.

Chapter 6: The Barrier That Wasn't

1 E. Mayr, "Weismann and Evolution," *Journal of the History of Biology* 18 (1985): 295–329, doi:10.1007/BF00138928.

2 P. J. Bowler, *The Eclipse of Darwinism: Anti-Darwinian Evolution Theories in the Decades around 1900* (Johns Hopkins Univ. Press, 1983).

3 M. L. Gabriel and S. Fogel, eds. *Great Experiments in Biology* (Prentice-Hall, 1955).

4 R. W. Burkhardt, "Closing the Door on Lord Morton's Mare: The Rise and Fall of Telegony," *Studies in the History of Biology* 3 (1979): 1–21.

5 A. Weismann, *Essays upon Heredity and Kindred Biological Problems* (Clarendon, 1891).

6 P. J. Bowler, "Theodor Eimer and Orthogenesis: Evolution by 'Definitely Directed Variation,'" *Journal of the History of Medicine and Allied Sciences* 34 (1979): 40–73, doi:10.1093/jhmas/XXXIV.1.40; J. R. Grehan and R. Ainsworth, "Orthogenesis and Evolution," *Systematic Biology* 34 (1985): 174–192, doi:10.2307/sysbio/34.2.174.

7 E. C. Case, "Cope: The man," *Copeia* 1940 (1940): 60–65; H. F. Osborn, *Cope: Master Naturalist. The Life and Letters of Edward Drinker Cope with a Bibliography of His Writings Classified by Subject* (Princeton Univ. Press, 1931).

8 A. S. Romer, "Cope versus Marsh," *Systematic Zoology* 13 (1964): 201–207; J. P. Davidson, *The Bone Sharp: The Life of Edward Drinker Cope* (Academy of Natural Sciences of Philadelphia, 1997).

9 R. T. Bakker, *The Dinosaur Heresies: New Theories Unlocking the Mystery of the Dinosaurs and Their Extinction* (Kensington, 2001).

10 S. M. Stanley, "An Explanation for Cope's Rule," *Evolution* 27 (1973): 1–26.

11 M. LaBarbera, "Analyzing Body Size as a Factor in Ecology and Evolution," *Annual Review of Ecology and Systematics* 20 (1989): 97–117.

12 D. W. E. Hone and M. J. Benton, "The Evolution of Large Size: How Does Cope's Rule Work?" *Trends in Ecology and Evolution* 20 (2005): 4–6, doi:http://dx.doi.org /10.1016/j.tree.2004.10.012.

13 E. D. Cope, *The Origin of the Fittest: Essays on Evolution* (Appleton, 1887).

14 August Weismann, *The Germ Plasm: A Theory of Heredity,* tr. W. Newton Parker and Harriet Rönnfeldt (Charles Scribner's Sons, 1893), 63.

Chapter 7: The Reverse Pinocchio

1 E. F. Keller, *A Feeling for the Organism: The Life and Work of Barbara McClintock* (W. H. Freeman, 1983); E. F. Keller, *The Century of the Gene* (Harvard Univ. Press, 2000).

2 B. Charlesworth, "The Evolution of Sex Chromosomes," *Science* 251 (1991): 1030–1033.

3 A. P. Kozlov, "Gene Competition and the Possible Evolutionary Role of Tumours," *Medical Hypotheses* 46 (1996): 81–84.

4 J. Sapp, "The Struggle for Authority in the Field of Heredity, 1900–1932: New Perspectives on the Rise of Genetics," *Journal of the History of Biology* 16 (1983): 311–342.

Chapter 8: A Multiplicity of Memory

1 W. L. Stone and E. A. Dratz, "Visual Photoreceptors," *Photochemistry and Photobiology* 26 (1977): 79–85.

2 R. V. Dippell, "The Development of Basal Bodies in *Paramecium,*" *Proceedings of the National Academy of Sciences* 61 (1968): 461–468; L. Margulis, "Centrioles and Kinetosomes in Animal Multicellularity," *Invertebrate Reproduction and Development* 23 (1993): 168–169, doi:10.1080/07924259.1993.9672311; M. J. Chapman, M. F. Dolan, and L. Margulis, "Centrioles and Kinetosomes: Form, Function, and Evolution," *Quarterly Review of Biology* 75 (2000): 1–20.

3 R. D. Manwell, *Introduction to Protozoology* (St. Martin's, 1961).

4 G. Vidal, "The Oldest Eukaryotic Cells," *Scientific American* 250 (1984): 48–57.

5 L. Margulis, M. Chapman, R. Guerrero and J. Hall, "The Last Eukaryotic Common Ancestor (LECA): Acquisition of Cytoskeletal Motility from Aerotolerant Spirochetes in the Proterozoic Eon," *Proceedings of the National Academy of Sciences* 103 (2006): 13080–13085, doi:10.1073/pnas.0604985103.

6 T. Fenchel and B. J. Finlay, "The Evolution of Life without Oxygen," *American Scientist* 82 (1994): 22–29.

7 P. A. Watson, "Function Follows Form: Generation of Intracellular Signals by Cell Deformation. *FASEB Journal* 5 (1991): 2013–2019.

8 Chapman and Margulis, "Centrioles and Kinetosomes"; M. J. Chapman, "One Hundred Years of Centrioles: The Henneguy-Lenhossek Theory, Meeting Report," *International Microbiology* 1 (1998): 233–236.

9 L. Margulis, Serial Endosymbiotic Theory (SET) and Composite Individuality:

Transition from Bacterial to Eukaryotic Genomes," *Microbiology Today* 31 (2004): 172–174.

10 L. Margulis and D. Sagan, *Acquiring Genomes: A Theory of the Origins of Species* (Basic Books, 2002).

11 T. Cavalier-Smith, "The Phagotrophic Origin of Eukaryotes and Phylogenetic Classification of Protozoa," *International Journal of Systematic and Evolutionary Microbiology* 52 (2002): 297–354, doi:10.1099/00207713–52–2-297.

12 R. Monastersky, "Oxygen Upheaval," *Science News* 142 (1992): 412–413; R. A. Berner, J. M. VanbedBrooks, and P. D. Ward, "Oxygen and Evolution," *Science* 316 (2007): 557–558.

13 Fenchel and Finlay, "Evolution of Life without Oxygen."

14 Margulis, "Centrioles and Kinetosomes"; Chapman and Margulis, "Centrioles and Kinetosomes"; Chapman, "One Hundred Years of Centrioles"; M. Chapman and M. C. Alliegro, "A Symbiogenetic Basis for the Centrosome?" *Symbiosis (Rehovot)* 44 (2007): 23–31; L. Margulis, M. F. Dolan, and R. Guerrero, "The Chimeric Eukaryote: Origin of the Nucleus from the Karyomastigont in Amitochondriate Protists," *Proceedings of the National Academy of Sciences* 97 (2000): 6954–6959, doi:10.1073/pnas.97.13.6954.

15 S. L. Bardy, S. Y. M. Ng, and K. F. Jarrell, "Prokaryotic Motility Structures," *Microbiology* 149 (2003): 295–304, doi:10.1099/mic.0.25948–0; M. J. Pallen, and N. J. Matzke, "From *The Origin of Species* to the Origin of Bacterial Flagella," *Nature Reviews Microbiology* 4 (2006): 784–790.

16 M. LaBarbera, "Why the Wheels Won't Go," *American Naturalist* 121 (1983): 395–408.

17 Dippell, "Development of Basal Bodies."

18 S. L. Bardy, S. Y. Ng, and K. F. Jarrell, "Prokaryotic Motility Structures," *Microbiology* 149 (2003): 295–304; C. Li, A. Motaleb, M. Sal, S. F. Goldstein, and N. W. Charon, "Spirochete Periplasmic Flagella and Motility," *Journal of Molecular and Microbiological Biotechnology* 2 (2000): 345–354; C. W. Wolgemuth, N. W. Charon, S. F. Goldstein, and R. E. Goldstein, "The Flagellar Cytoskeleton of the Spirochetes," *Journal of Molecular Microbiology and Biotechnology* 11 (2006): 221–227.

19 G. Dubinina, M. Grabovich, N. Leshcheva, F. A. Rainey, and E. Gavrish, "*Spirochaeta perfilievii* sp. nov., an Oxygen-Tolerant, Sulfide-Oxidizing, Sulfur-and Thiosulfate-Reducing Spirochaete Isolated from a Saline Spring," *International Journal of Systematic and Evolutionary Microbiology* 61 (2011): 110–117, doi:10.1099/ijs.0.018333–0; L. J. Stal, "Coastal Microbial Mats: The Physiology of a Small-Scale Ecosystem," *South African Journal of Botany* 67 (2001): 399–410; E. A. Stephens, O. Braissant, and P. T. Visscher, "Spirochetes and Salt Marsh Microbial Mat Geochemistry: Implications for the Fossil Record," *Carnet de Géologie/ Notebooks on Geology*, Brest, article 2008/09 (2008).

20 Margulis et al., "Last Eukaryotic Common Ancestor."

21 L. Margulis, M. F. Dolan, and J. H. Whiteside, "'Imperfections and Oddities' in the Origin of the Nucleus," *Paleobiology* 31 (2005): 175–191, doi:10.1666/00948373(2005)031[0175:iaoito]2.0.co;2.

22 Li et al., "Spirochete Periplasmic Flagella and Motility."

23 N. Yubuki and B. S. Leander, "Evolution of Microtubule Organizing Centers across the Tree of Eukaryotes," *Plant Journal* 75 (2013): 230–244, doi:10.1111/tpj.12145.

24 Bardy et al., "Prokaryotic Motility Structures"; R. Lux, A. Moter, and W. Shi, "Chemotaxis in Pathogenic Spirochetes: Directed Movement toward Targeting Tissues?" *Journal of Molecular and Microbiological Biotechnology* 2 (2000): 355–364.

25 L. Margulis, "The Conscious Cell," *Annals of the New York Academy of Sciences* 929 (2001): 55–70, doi:10.1111/j.1749–6632.2001.tb05707.x.

26 J. S. Turner, *The Tinkerer's Accomplice: How Design Emerges from Life Itself* (Harvard Univ. Press, 2007); G. G. Gundersen and T. A. Cook, "Microtubules and Signal Transduction," *Current Opinion in Cell Biology* 11 (1999): 81–94, doi:http://dx.doi.org/10.1016/S0955–0674(99)80010–6; P. Nick, "Microtubules, Signalling and Abiotic Stress," *Plant Journal* 75 (2013): 309–323, doi:10.1111/tpj.12102; K. C. Vaughn and J. D. I. Harper, "Microtubule-Organizing Centers and Nucleating Sites in Land Plants," *International Review of Cytology* 181 (1998): 75–149.

Chapter 9: One Is the Friendliest Number

1 J. Dupré and M. A. O'Malley, "Varieties of Living Things: Life at the Intersection of Lineage and Metabolism," *Philosophy and Theory in Biology* 1 (2009): 1–25, doi:http://dx.doi.org/10.3998/ptb.6959004.0001.003; J. S. Turner, "A Superorganism's Fuzzy Boundary," *Natural History* 111 (2002): 62–67; R. E. Ulanowicz, "The Organic in Ecology," *Ludus Vitalis* 9 (2001): 183–204.

2 M. F. Glaessner, "The Emergence of Metazoa in the Early History of Life," *Precambrian Research* 20 (1983): 427–441; B. Balavoine and A. Adoutte, "One or Three Cambrian Radiations?" *Science* 280 (1998): 397–398; D. Erwin, J. Valentine, and D. Jablonski, "The Origin of Animal Body Plans," *American Scientist* 85 (1997): 126–137.

3 D. J. Des Marais, "Microbial Mats and the Early Evolution of Life," *Trends in Ecology and Evolution* 5 (1990): 140–144; Y. Watanabe, J. E. J. Martini, and H. Ohmoto, "Geochemical Evidence for Terrestrial Ecosystems 2.6 Billion Years Ago," *Nature* 408 (2000): 574–578.

4 L. J. Stal, "Coastal Microbial Mats: The Physiology of a Small-Scale Ecosystem," *South African Journal of Botany* 67 (2001): 399–410.

5 B. Hölldobler and E. O. Wilson, *The Superorganism: The Beauty, Elegance, and Strangeness of Insect Societies* (W. W. Norton, 2009).

6 Dupré and O'Malley, "Varieties of Living Things"; J. Xu and J. I. Gordon, "Honor Thy Symbionts," *Proceedings of the National Academy of Sciences* 100 (2003): 10452–10459, doi:10.1073/pnas.1734063100.

7 J. Maynard Smith and E. Szathmáry, *The Major Transitions in Evolution* (W. H. Freeman/Spektrum, 1995).

8 E. O. Wilson, *Sociobiology: The New Synthesis* (Belknap/Harvard Univ. Press, 1975).

9 D. C. Queller, "W. D. Hamilton and the Evolution of Sociality," *Behavioral Ecology* 12 (2001): 261–264, doi:10.1093/beheco/12.3.261-a.

10 U. Segerstrale, *Nature's Oracle: The Life and Work of W. D. Hamilton* (Oxford Univ. Press, 2013).

11 W. D. Hamilton, *Narrow Roads of Gene Land: Vol. 1. Evolution of Social Behavior* (W. H. Freeman, 1996), 4.

12 A. Grafen, "William Donald Hamilton. 1 August 1936–7 March 2000. Elected FRS 1980," *Biographical Memoirs of the Fellows of the Royal Society* 50 (2004): 109–132.

13 E. O. Wilson, *The Insect Societies* (Belknap/Harvard Univ. Press, 1971).

14 F. R. Prete, "The Conundrum of the Honey Bees: One Impediment to the Publication of Darwin's Theory," *Journal of the History of Biology* 23 (1990): 271–290, doi:10.2307/4331130.

15 Prete, "Conundrum of the Honey Bees."

16 B. L. Thorne, Evolution of Eusociality in Termites," *Annual Review of Ecology and Systematics* 28 (1997): 27–54, doi:10.2307/2952485.

17 J. S. Turner, "Termites as Models of Swarm Cognition," *Swarm Intelligence* 5 (2011): 19–43, doi:10.1007/s11721–010–0049–1.

18 W. D. Hamilton, "Altruism and Related Phenomena, Mainly in Social Insects," *Annual Review of Ecology and Systematics* 3 (1972): 193–232, doi:10.2307/2096847.

19 H. Kirby, L. Margulis, and M. Yamin, "Harold Kirby's Symbionts of Termites: Karyomastigont Reproduction and Calonymphid Taxonomy," *Symbiosis (Rehovot)* 16 (1994): 7–63; W. A. Sands, "The Association of Termites and Fungi," in *Biology of Termites*, Vol. I, ed. K. Krishna and F. M. Weesner (Academic Press, 1969), 495–524.

20 Thorne, "Evolution of Eusociality in Termites"; P. W. Sherman, E. A. Lacey, H. K. Reeve, and L. Keller, "The Eusociality Continuum," *Behavioral Ecology* 6 (1994): 102–108.

21 L. M. Reilly, "Measurements of Inbreeding and Average Relatedness in a Termite Population," *American Naturalist* 130 (1987): 339–349, doi:10.2307/2461889.

22 J. P. E. C. Darlington, "The Structure of Mature Mounds of the Termite *Macrotermes michaelseni* in Kenya," *Insect Science and Its Application* 6 (1985): 149–156; and W. V. Harris, "Termite Mound Building," *Insectes Sociaux* 3 (1956): 261–268.

23 For more on the discussion that follows about the extended organism concept, see J. S. Turner, *The Extended Organism: The Physiology of Animal-Built Structures* (Harvard Univ. Press, 2000); and J. S. Turner, "Homeostasis and the Physiological Dimension of Niche Construction Theory in Ecology and Evolution," *Ecology and Evolution* 30 (2016): 203–219, doi:10.1007/s10682–015–9795–2.

24 J. E. Lovelock, *Gaia: A New Look at Life on Earth* (Oxford Univ. Press, 1987); L. Margulis and D. Sagan, eds., *Slanted Truths: Essays on Gaia, Symbiosis, and Evolution* (Copernicus, 1997).

Chapter 10: The Hand of Whatever

1 F. Macrae, "Scientist Accused of Playing God After Creating Artificial Life by Making Designer Microbe from Scratch—But Could It Wipe Out Humanity?" *The Daily Mail (UK)*, 3 June 2010.

2 F. Dyson, *Origins of Life*, 2nd ed. (Cambridge Univ. Press, 1999); D. Rohlfing, "Coacervate-like Microspheres from Lysine-Rich Proteinoid," *Origins of Life* 6 (1975): 203–209, doi:10.1007/BF01372406; J. W. Schopf, "The Evolution of the Earliest Cells," *Scientific American* 239 (1978): 110–138; J. W. Schopf, "Microfossils of the Early Archean Apex Chert: New Evidence of the Antiquity of Life," *Science* 260 (1993): 640–646; C. R. Woese, *Carolina Biology Readers*, Vol. 13 (Carolina Biological Supply, 1984), 1–31.

3 K. A. Maher and D. J. Stevenson, "Impact Frustration of the Origin of Life," *Nature* 331 (1988): 612–614.

4 R. Shapiro, *Origins: A Skeptic's Guide to the Creation of Life on Earth* (Heinemann, 1986); I. Fry, *The Emergence of Life on Earth: A Historical and Scientific Overview* (Rutgers Univ. Press, 2000).

5 See, for example, the selected titles and internet resources listed at Library of Congress, Science Reference Services, Science Reference Guides, "The Origins of Life & the Universe: A Guide to Selected Resources," https://www.loc.gov/rr /scitech/SciRefGuides/originsoflife.html (22 July 2015).

6 J. Strick, "Spontaneous Generation," in *Encyclopedia of Microbiology*, Vol. 4, 2nd ed. (Elsevier, 2000), 364–376; J. E. Strick, *Sparks of Life: Darwinism and the Victorian Debates over Spontaneous Generation* (Harvard Univ. Press, 2000).

7 Maher and Stevenson, "Impact Frustration"; C. M. Fedo and M. J. Whitehouse, "Metasomatic Origin of Quartz-Pyroxene Rock, Akilia, Greenland, and Implications for Earth's Earliest Life," *Science* 296 (2002): 1448.

8 E. Szathmary, "The Evolution of Replicators," *Philosophical Transactions of the Royal Society of London Series B* 355 (2000): 1669–1676.

9 T. R. Cech, "RNA Finds a Simpler Way," *Nature* 428 (2004): 263–264; M. M. Waldrop, "Finding RNA Makes Proteins Gives 'RNA World' a Big Boost," *Science* 256 (1992): 1396–1397.

10 Szathmary, "Evolution of Replicators."

11 Cech, "RNA Finds a Simpler Way"; Waldrop, "Finding RNA."

12 R. Shapiro, "A Replicator Was Not Involved in the Origin of Life," *IUBMB Life* 49 (2000): 173–176, doi:10.1080/152165400306160.

13 A. Cairns-Smith and H. Hartman, eds., *Clay Minerals and the Origin of Life* (Cambridge Univ. Press, 1986); A. G. Cairns-Smith, *Genetic Takeover and the Mineral Origins of Life* (Cambridge Univ. Press, 1982); A. G. Cairns-Smith, *Seven Clues to the Origin of Life: A Scientific Detective Story* (Cambridge Univ. Press, 1986).

14 M. Sernetz, B. Gelleri, and J. Hoffman, "The Organism as Bioreactor: Interpretation of the Reduction Law of Metabolism in Terms of Heterogeneous Catalysts and Fractal Structure," *Journal of Theoretical Biology* 117 (1985): 209–230.

15 Cairns-Smith, *Clay Minerals*, and Cairns-Smith, *Genetic Takeover*.

16 G. Wächtershäuser, "Before Enzymes and Templates: Theory of Surface Metabolism," *Microbiological Reviews* 52 (1988): 452–484.

17 V. J. A. Novak, "Present State of the Coacervate-in-Coacervate Theory: Origin and Evolution of Cell Structure," *Origins of Life* 14 (1984): 513–522, doi:10.1007/BF00933699.

18 Rohlfing, "Coacervate-like Microspheres"; Novak, "Present State."

19 B. Ecanow, "Drug Delivery Compositions and Methods," US 4963367A, Patents, 16 October 1990, https://www.google.com/patents/US4963367.

20 S. W. Fox, "A Theory of Macromolecular and Cellular Origins," *Nature* 205 (1965): 328–340; S. W. Fox, "Metabolic Microspheres," *Naturwissenschaften* 67 (1980): 378–383.

21 Shapiro, *Origins*.

22 Fox, "Metabolic Microspheres."

23 Fox, "Theory of Macromolecular and Cellular Origins."

24 Szathmary, "Evolution of Replicators"; B. G. Bag and V. Kiedrowski, "Templates, Autocatalysis and Molecular Replication," *Pure and Applied Chemistry* 68 (1996): 2145–2152.

25 J. Reineke, S. Tenzer, M. Rupnik, A. Koschinski, O. Hasselmayer, A. Schrattenholz, H. Schild, and C. von Eichel-Streiber, "Autocatalytic Cleavage of *Clostridium difficile* Toxin B," *Nature* 446 (2007): 415–419, doi:10.1038/nature05622.

26 Wächtershäuser, "Before Enzymes and Templates."

27 R. E. Morel and G. Fleck, "A Fourth Law of Thermodynamics," *Chemistry* 15 (2006): 305–310.

28 R. Swenson, "The Fourth Law of Thermodynamics or the Law of Maximum Entropy Production," *Chemistry* 18 (2009): 333–339.

29 Sernetz et al., "Organism as Bioreactor."

30 W. A. Hermann, "Quantifying Global Exergy Resources," *Energy* 31 (2006): 1685–1702, doi:http://dx.doi.org/10.1016/j.energy.2005.09.006.

31 S. A. Kauffman, *The Origins of Order: Self-Organization and Selection in Evolution* (Oxford Univ. Press, 1993).

32 J. S. Turner, "Homeostasis and the Physiological Dimension of Niche Construction Theory in Ecology and Evolution," *Ecology and Evolution* 30 (2016): 203–219, doi:10.1007/s10682–015–9795–2.

33 Maher and Stevenson, "Impact Frustration."

34 F. Gauthier-Lafaye, P. Holliger, and P.-L. Blanc, "Natural Fission Reactors in the Franceville Basin, Gabon: A Review of the Conditions and Results of a 'Critical Event' in a Geologic System," *Geochimica et Cosmochimica Acta* 60 (1996): 4831–4852.

35 L. Margulis, "The Conscious Cell," *Annals of the New York Academy of Sciences* 929 (2001): 55–70, doi:10.1111/j.1749–6632.2001.tb05707.x.

Chapter 11: Plato Street

1 F. B. Golley, *A History of the Ecosystem Concept in Ecology* (Yale Univ. Press, 1993).

2 N. G. Slack, *Evelyn Hutchinson and the Invention of Modern Ecology* (Yale Univ. Press, 2011); T. E. Lovejoy, "George Evelyn Hutchinson. 13 January 1903–17 May 1991," *Biographical Memoirs of Fellows of the Royal Society* 57 (2011): 167–177, doi:10.1098 /rsbm.2010.0016.

3 Lovejoy, "George Evelyn Hutchinson"; L. B. Slobodkin and N. G. Slack, "George Evelyn Hutchinson: 20th Century Ecologist," *Endeavour* 23 (1999): 24–30.

4 R. E. Cook, "Raymond Lindeman and the Trophic-Dynamic Concept in Ecology," *Science* 198 (1977): 22–26; C. B. Reif, "Memories of Raymond Laurel Lindeman," *Bulletin of the Ecological Society of America* 67 (1986): 20–25, doi:10.2307/20166483.

5 R. L. Lindeman, "The Trophic-Dynamic Aspect of Ecology," *Ecology* 23 (1942): 399–417, doi:10.2307/1930126.

6 G. E. Hutchinson, "Concluding Remarks," *Cold Spring Harbor Symposia on Quantitative Biology* 22 (1957): 415–427, doi:10.1101/SQB.1957.022.01.039.

7 R. Lewontin, "Gene, Organism and Environment," in *Evolution from Molecules to Men*, ed. D. S. Bendall (Cambridge Univ. Press, 1983), 273–286; R. C. Lewontin, *The Triple Helix: Gene, Organism and Environment* (Harvard Univ. Press, 2000).

8 R. B. Huey and R. D. Stevenson, "Integrating Thermal Physiology and Ecology of Ectotherms: A Discussion of Approaches," *American Zoologist* 19 (1979): 357–366; R. D. Stevenson, C. R. Peterson, and J. S. Tsugi, "The Thermal Dependence of Locomotion, Tongue Flicking, Digestion and Oxygen Consumption in the Wandering Garter Snake," *Physiological Zoology* 58 (1985): 46–57; F. H. Pough, "Organismal Performance and Darwinian Fitness: Approaches and Interpretations," *Physiological Zoology* 62 (1989): 199–236.

9 P. E. Hertz, R. B. Huey, and E. Nevo, "Homage to Santa Anita: Thermal Sensitivity of Sprint Speed in Agamid Lizards," *Evolution* 37 (1983): 1075–1084.

10 M. L. Cody and J. M. Diamond, eds., *Ecology and Evolution of Communities* (Belknap/Harvard Univ. Press, 1975), viii.

11 C. E. Cooper, "Myrmecobius fasciatus (Dasyuromorphia: Myrmecobiidae)," *Mammalian Species* 43 (2011): 129–140.; H. J. de Knegt, et al., "The Spatial Scaling of Habitat Selection by African Elephants," *Journal of Animal Ecology* 80 (2011): 270–281, doi:10.1111/j.1365–2656.2010.01764.x; H. F. Hirth, "The Ecology of Two Lizards on a Tropical Beach," *Ecological Monographs* 33 (1963): 83–111; T. E. Lacher, "Wing Morphology, Flights Speeds and Insights into Niche Structure in Caribbean Bats from Dominica," *Chiroptera Neotropical* 18 (2012): 1067–1073; G. Jones and M. W. Holderied, "Bat Echolocation Calls: Adaptation and Convergent Evolution," *Proceedings of the Royal Society B: Biological Sciences* 274 (2007): 905–912, doi:10.1098 /rspb.2006.0200; J. Rydell and J. R. Speakman, "Evolution of Nocturnality in Bats: Potential Competitors and Predators during Their Early History," *Biological Journal of the Linnean Society* 54 (1995): 183–191, doi:10.1111/j.1095–8312.1995.tb01031.x; C. A. Simon and G. A. Middendorf, "Resource Partitioning by an Iguanid Lizard: Temporal and Microhabitat Aspects," *Ecology* 57 (1976): 1317–1320; S. Turkarslan, D. J. Reiss, G. Gibbins, W. L. Su, M. Pan, J. C. Bare, C. L. Plaisier, and N. S. Baliga, "Niche Adaptation by Expansion and Reprogramming of General Transcription Factors," *Molecular Systems Biology* 7 (2011): doi:http://www.nature.com/msb /journal/v7/n1/suppinfo/msb201187_S1.html.

12 Jones and Holderied, "Bat Echolocation Calls"; Rydell and Speakman, Evolution of Nocturnality in Bats"; J. H. Fullard, R. M. R. Barclay, and D. W. Thomas, "Echolocation in Free-Flying Atiu Swiftlets (*Aerodramus sawtelli*)," *Biotropica* 25 (1993): 334–339, doi:10.2307/2388791; J. R. Speakman, "The Evolution of Flight and Echolocation in Bats: Another Leap in the Dark," *Mammal Review* 31 (2001): 111–130, doi:10.1046/j.1365–2907.2001.00082.x.

13 J. S. Turner, "Extended Phenotypes and Extended Organisms," *Biology and Philosophy* 19 (2004): 327–352; J. S. Turner, *The Tinkerer's Accomplice: How Design Emerges from Life Itself* (Harvard Univ. Press, 2007); J. S. Turner, "Homeostasis and the Physiological Dimension of Niche Construction Theory in Ecology and Evolution," *Ecology and Evolution* 30 (2016): 203–219, doi:10.1007/s10682 –015–9795–2.

14 K. Laland et al. "Does Evolutionary Theory Need a Rethink? *Nature* 514 (2014): 161–164; K. N. Laland, J. Odling-Smee, and M. W. Feldman, "Niche Construction, Biological Evolution and Cultural Change," *Behavioral and Brain Sciences* 23 (2000): 131–175; F. J. Odling-Smee, K. N. Laland, and M. W. Feldman, "Niche Construction," *American Naturalist* 147 (1996): 641–648.

15 K. Baumann, "Stem Cells: Self-Help in the Niche," *Nature Reviews Molecular Cell Biology* 13 (2012): 61, doi:10.1038/nrm3279; J. Odling-Smee and J. S. Turner, "Niche Construction Theory and Human Architecture," *Biological Theory* 6 (2011): 283–289, doi:10.1007/s13752–012–0029–3; K. N. Laland, J. Odling-Smee, and S.

Myles, "How Culture Shaped the Human Genome," *Nature Reviews Genetics* 11 (2010): 137–148.

16 Turner, "Homeostasis and Physiological Dimension."

17 T. Patterson, CNN, "Who's Really Flying the Plane?," http://www.cnn.com /2012/03/24/travel/autopilot-airlines/, updated 26 March 2012.

18 A. Feduccia, "Explosive Evolution in Tertiary Birds and Mammals," *Science* 267 (1995): 637–638.

19 R. O. Prum, "Development and Evolutionary Origin of Feathers," *Journal of Experimental Zoology* 285 (1999): 291–306; P. J. Regal, "The Evolutionary Origin of Feathers," *Quarterly Review of Biology* 50 (1975): 35–66.

20 W. J. Bock, "The Arboreal Origin of Avian Flight," in *The Origin of Birds and the Evolution of Flight,* ed. Kevin Padian (California Academy of Sciences, 1986), 57–72; J. H. Ostrom, "The Cursorial Origin of Avian Flight," in ibid., 73–81.

21 J. W. Bahlman, S. M. Swartz, D. K. Riskin, and K. S. Breuer, "Glide Performance and Aerodynamics of Non-Equilibrium Glides in Northern Flying Squirrels (*Glaucomys sabrinus*)," *Journal of the Royal Society Interface* 10 (2013), doi:10.1098 /rsif.2012.0794.

22 Prum, "Development and Evolutionary Origin of Feathers"; P. F. A. Maderson and L. Alibardi, "The Development of the Sauropsid Integument: A Contribution to the Problem of the Origin and Evolution of Feathers," *American Zoologist* 40 (2000): 513–529, doi:10.1093/icb/40.4.513.

Epilogue: Evolution, Purpose, and Desire

1 Described vividly in Stephen Jay Gould's essay "A Visit to Dayton," *Natural History* 90 (1981): 8; S. J. Gould, *Hen's Teeth and Horse's Toes* (Penguin, 1990).

2 E. J. Larson, *Summer for the Gods: The Scopes Trial and America's Continuing Debate over Science and Religion* (Basic Books, 2008).

3 M. Matsumura and L. Mead, National Center for Science Education, "'Ten Major Court Cases about Evolution and Creationism," https://ncse.com /library-resource/ten-major-court-cases-evolution-creationism.

4 See U.S. National Bioethics Advisory Commission, "Ethical Issues in Human Stem Cell Research," 1999, National Bioethics Advisory Commission Publications, https://bioethicsarchive.georgetown.edu/nbac/pubs.html.

5 C. Dean, "Believing in Scripture but Playing by Science's Rules," *New York Times,* 12 February 2007.

6 T. Nagel, *Mind and Cosmos: Why the Materialist Neo-Darwinian Conception of Nature Is Almost Certainly False* (Oxford Univ. Press, 2012).

CREDITS

Grateful acknowledgment is given to the artists and publishers below for the use of their images in *Purpose and Desire*.

Figure 2.1   (*left*) Eugene Marais; (*right*) J. Scott Turner

Figure 2.2   J. Scott Turner

Figure 3.1   Image in the public domain.

Figure 3.2   Images in the public domain.

Figure 4.1   Image in the public domain.

Figure 4.2   J. Scott Turner

Figure 4.3   Advertisements from *Scientific American* 196 (January 1957). Permission granted by Servo Corporation of America.

Figure 4.4   J. Scott Turner

Figure 4.5   J. Scott Turner

Figure 4.6   Adapted by J. Scott Turner from K. Nagashima et al., "Neuronal Circuitries Involved in Thermoregulation," *Autonomic Neuroscience* 85 (2000), permission granted by Elsevier.

Figure 5.1   Images in the public domain.

Figure 5.2   J. Scott Turner

Figure 5.3   J. Scott Turner

Figure 5.4   Images in the public domain.

Figure 6.1   Image in the public domain.

Figure 6.2   J. Scott Turner

Figure 6.3   J. Scott Turner

Figure 6.4   J. Scott Turner

Figure 6.5   J. Scott Turner

Figure 6.6   Image in the public domain.

Figure 6.7   J. Scott Turner

Figure 7.1   (*left*) J. Scott Turner; (*right*) Image in the public domain.

Figure 7.2   Image in the public domain.

Figure 7.3   J. Scott Turner

Figure 7.4   Image in the public domain.

Figure 8.1   J. Scott Turner

Figure 8.2   J. Scott Turner

Figure 8.3   Image in the public domain.

Figure 8.4   J. Scott Turner

Figure 8.5   J. Scott Turner

Figure 8.6   J. Scott Turner

Figure 8.7   Rendering by Katherine Delisle, from L. Margulis et al., "Imperfections and oddities in the Origin of the Nucleus," *Paleobiology* 31 (2005): 175–91.

Figure 9.1   Photograph taken in March 1990 at Nagoya University, Japan, © Tokyo Zoological Park Society.

Figure 9.2   Image in the public domain.

Figure 9.3   Figure by The Company of Biologists, from K. Rafiq et al., "The Genomic Regulatory Control of Skeletal Morphogenesis in the Sea Urchin," *Development* 139, no. 3 (2012): 579–90.

Figure 9.4   J. Scott Turner

Figure 9.5   J. Scott Turner

Figure 9.6   J. Scott Turner

Figure 10.1   J. Scott Turner

Figure 10.2   J. Scott Turner

Figure 10.3   J. Scott Turner

Figure 10.4   J. Scott Turner

Figure 11.1   J. Scott Turner

Figure 11.2   J. Scott Turner

Figure 11.3   J. Scott Turner

Figure 11.4   J. Scott Turner

Figure 11.5   G. Evelyn Hutchinson Papers (MS 649). Manuscripts and Archives, Yale University Library.

Figure 11.6   Y. H. Edmondson's dedication to *Limnology and Oceanography* 16, no. 2 (1971). Permission granted by John Wiley & Sons.

Figure 11.7   Adapted by J. Scott Turner from P. E. Hertz et al., "Homage to Santa Anita," *Evolution* 37 (1983): 1075–84.

Figure 11.8   J. Scott Turner

Epilogue   J. Scott Turner

INDEX